A Case Study for Computer Ethics in Context

Aimed at addressing the difficulties associated with teaching often abstract elements of technical ethics, this book is an extended fictional case study into the complexities of technology and social structures in complex organisations. Within this case study, an accidental discovery reveals that the algorithms of Professor John Blackbriar are not quite what they were purported to be. Over the course of 14 newspaper articles, a nebula of professional malpractice and ethical compromise is revealed, ultimately destroying the career of a prominent, successful academic.

The case study touches on many topics relevant to ethics and professional conduct in computer science, and on the social structures within which computer science functions. Themes range from the growing influence of generative AI to the difficulties in explaining complex technical processes to a general audience, also touching on the environmental consequences of blockchain technology and the disproportionate gender impacts of Coronavirus. Each new revelation in the case study unveils further layers of complexity and compromise, leading to new technical and social issues that need to be addressed.

Directly aimed at making ethics in the digital age accessible through the use of real-world examples, this book appeals to computer science students at all levels of the educational system, as well as making an excellent accompaniment to lecturers and course convenors alike.

Dr Michael James Heron is a senior lecturer in interaction design and has been working with ethics in computer science and in gaming for pretty much his entire career. He has seen some things.

Pauline Helen Belford is a teacher in interaction design and games and has been involved in the educational system at all levels of the curriculum for two decades. She too has seen some things.

A Case Study for Computer Ethics in Context

The Scandal in Academia

Michael James Heron and Pauline Helen Belford

CRC Press is an imprint of the
Taylor & Francis Group, an **informa** business

First edition published 2024
by CRC Press
2385 NW Executive Center Drive, Suite 320, Boca Raton FL 33431

and by CRC Press
4 Park Square, Milton Park, Abingdon, Oxon, OX14 4RN

CRC Press is an imprint of Taylor & Francis Group, LLC

ISBN: 9781032546902 (hbk)
ISBN: 9781032546919 (pbk)
ISBN: 9781003426172 (ebk)

DOI: 10.1201/9781003426172

Typeset in Minion Pro
by KnowledgeWorks Global Ltd.

Contents

CHAPTER 1

About This Case Study

Ethical instruction in computing science courses is an important part of organisational credibility and professional body accreditation. It serves an important role in anchoring student skillsets within the context of societal norms, while also helping future professionals understand the implications of what they do. The British Computer Society (BCS) emphasises that students must be exposed to professional and ethical outlooks before BCS accreditation can be awarded. The joint Association of Computer Machinery (ACM) and IEEE curriculum guidelines for undergraduate degree programmes likewise stress the importance of ethical and professional understanding. A framework for incorporating this kind of content into courses is important, and while these guidelines offer support for the importance of the topic, there remains a need for a coherent set of instructional material that helps contextualise the issues and relate them to each other. Much of the difficulty associated with meaningful ethical instruction is complicated by the emotional distance that students can put between themselves and the topic (Heron & Belford, 2015) and the lack of cohesive examples drawn from real life only exacerbates the difficulty of establishing relevance.

Back in the nineties, a book was published under the name *The Case of the Killer Robot: Stories about the Professional, Ethical and Societal Dimensions of Computing*. The book was an extension of papers published in the ACM Computers and Society (SIGCAS) newsletters (Epstein, 1994a, 1994b). The Case of the Killer Robot was a genuinely innovative, original, and entertaining fictionalised account of a software development company beset with institutional problems and faced with a growing legal and reputational crisis. The story begins with the death of a factory

DOI: 10.1201/9781003426172-1

worker at the hands of an industrial robot developed and built by the company that serves as the key player in the unfolding revelations. It tells the story through newspaper articles that reveal new elements of the company, people, and management structures within which the deadly robot had been constructed. It all starts with a single malfunctioning piece of code in a subroutine of the robot's systems. The programmer who wrote the code is then apprehended for the crime – the prosecuting attorney for the dead man's family throwing him to the judgement of the court of human opinion assisted by a newspaper called the 'Silicon Observer'. That in turn may or may not have had an axe to grind in the presentation of the case. As the story unfolds, we find out about the organisational problems in the company that built the software. We see incidences of faked testing. Of internal power struggles. We encounter dubious management paradigms, plagiarism of source code, and more. The end result of the case study is to underscore both the difficulty of identifying where things go wrong and outlining how individuals should have acted under the circumstances. Blame is difficult to apportion because everyone shares some measure of complicity. Much can be ascertained about readers of the study by where they place the largest burden of responsibility. In the end, everyone is to blame and no one is to blame, but this judgement only emerges through consideration of the chain of cause and effect and of the relationships between each of the characters in the story.

The authors of this book have been fans of this case study for a very long time. We have individually or collectively used it whenever an opportunity to discuss ethics in computing has arisen. It is sometimes discussed over a single lecture. More often, it is covered in depth over multiple weeks and with multiple groups of students each with their own opinions and thoughts about the activities encountered. Over the years, we have had cause to appreciate the depth of the study as demonstrated through the variation of the viewpoints expressed by those to whom we have presented it.

The Case of the Killer Robot is very old now, and in a field that moves as quickly as computing it couldn't possibly fail to show its age. It talks of no longer existent controversies in project management methodologies. It has arguments where the proponents are fighting over an issue that has long been put to bed. One discussion focuses on prototyping and the role of the user, as if it is in any way controversial in modern software development. Its discussion of user interfaces is anachronistic and the organisation complexities it raises are products of a very different time. As a case study, it's still very valid, very interesting, and well worth reading. However, an educator must spend a lot of time talking around the historical quirks of the study if it is to be used to illustrate issues of modern ethics. In effect, the footnotes overwhelm the content.

And thus, we introduce the Scandal in Academia.

This book owes an enormous intellectual debt to Richard Epstein's original case study. The Scandal in Academia is presented not as a replacement to the Case of the Killer Robot, but as a complement. It provides a fully integrated ethical case study in the same style but chooses a different scenario and introduces different issues, many of which were not part of the Killer Robot scenario. In 1994, few people would have dreamed of how social networks would change the way in which journalism worked, or how controlling access to information online would be a Sisyphean task with little hope of success. Accessibility of information and discrimination in the workplace were issues that didn't have the prominence they do now. These elements and more are incorporated into the Scandal in Academia as major issues for consideration and debate.

It's not our intent within this text to paint a cartoonish picture of 'right' versus 'wrong'.

We will not attempt to proselytise a particular ethical position. This case study is used as a jumping off point for discussion of the surrounding issues. An illustrative discussion of how we might draw in the complexity of the real world is provided after each of the case study articles. No claim is made that this discussion is definitive or even necessarily comprehensive. It's just an example of where the discussion could go. We expect, and encourage, users and students to disagree and to come up with counterexamples. This is, indeed, why a case study of this nature is such a useful pedagogical tool.

The case study may be used in many ways, and the authors of this chapter have profitably employed both this study and the Killer Robot study which inspired it. However, key to the study is the layering of revelation – each article needs some time to 'sink in', and each requires discussion and consideration of how the perceived responsibility of each player manifests over time. First, readers must identify the ethical issues that are raised, and then how these may have contributed to the emerging story at the heart of the scenario. These issues, once identified, become discussion points for individual or group consideration. We provide a full outline of how to employ this study in Chapter 2. In such classes, what is most valuable is the discussion – in the best cases, classroom engagement will reveal moral and ethical points that are perfectly valid and yet things that we as authors had never considered. Importantly though, this book is not intended to put forward our view, but instead to give common ground for issues to be discussed and then gradually nuanced.

A significant disclaimer is needed here before the articles are presented. Almost all of the issues discussed in this case study have been inspired by real life. Some of them we read about, some of them we directly experienced, and some were experience by others we know. The incident outlined herein is an entirely fictionalised composite of these elements, and those elements have been reshaped and restructured to support our case study. This case study does not reflect any organisation with which we may have been associated. The characters involved may occasionally share personal traits with individuals with whom we are acquainted, but they are not stand-ins or composites of anyone in particular. As the old disclaimer goes, 'any resemblance to persons living or dead is entirely coincidental'.

MAP OF THE SCANDAL

The Scandal in Academia is large, with the core text consisting of 14 separate articles which should be read sequentially. There are also several additional articles which provide additional perspectives that are not core to the narrative. Each of these core articles introduces a new set of ethical issues or reinforces issues previously introduced. Each layers further complexity onto the core story and introduces new pressures and incentives for the behaviours of everyone involved. It is expected that reader perceptions of all participants will change as new information is made available. One of the core aspects of this case study though is that our entire view of the scenario comes through one source – the local newspaper for the city. As such, even the information we are presented is not necessarily completely accurate, complete, or unbiased. A common thread throughout the study is verifying and evaluating the information we are presented, and readers should be encouraged throughout to question the motivations of those who are presenting us with the information. That includes us, as the authors.

Article One – A Scandal in Academia

In this chapter we learn about Professor Blackbriar and his suspension from the fictional University of Dunglen. He is alleged to have participated in scientific misconduct, and the initial newspaper article includes comments from the principal and others about the role of ethics in professional life. This serves as the introduction to the case study, and an opportunity for educators to discuss whatever ethical codes and moral foundations they feel are appropriate within the context of their courses. Concentrating on a fixed code such as the BCS Code of Conduct or the ACM Code of Ethics would be a useful way of anchoring the discussions to follow, but the material can also be productively used in an agnostic fashion.

Article Two – Student Suspensions at Scandal-Ridden University

In this chapter we find out that two of the professor's postgraduate students have been suspended, and we learn a little more about the commercial interests that may have exacerbated the situation in which the scientific misconduct may have occurred. Ethical issues introduced here include corporate interests in academia; reputational damage; transparency of artificial intelligence; the environmental impact of blockchain technologies; and the issue of reproducibility in science.

Article Three – Multimillion-Pound Consequences for Research Fiddle

In this chapter we get to know some of the context of the professor's work, and the likely financial consequences if the allegations are true. We find out that his research has been used in several important projects, and that there are many millions of pounds on the line if his research does not produce the promised results. Here we introduce the issues of corporate influence in academia and the impact of academia on industry; speculative investment on the basis of uncorroborated research; how Coronavirus and other international incidents can impact on just-in-time economies; and the mental biases that impact on good decision-making.

Article Four – Students Speak Out

In this chapter we get to hear the story from the suspended students. They tell a tale of academic serfdom and the power of social contexts. They discuss the difficulties in keeping an important colleague happy, and the impact that might have on their eventual future success or failure. Ethical issues introduced here include workplace relations; power differentials; the role of the postgraduate student in modern academia; closed data sets and the publication imperative; sexism in the workplace; large language models and AI authoring; workplace bullying; and issues of accessibility of information resources.

Article Five – Leaked Minute Lays Bear University Culture

In this chapter we find out about a minute leaked from the principal's office in which he tells all staff members that if they cannot acquire funding for their projects then they and their dependent staff would be dismissed. We also find out the impact that research assessment exercises have had on the university's past and present league table standings. Issues introduced here include the ethics of whistleblowing; the impact of quantification in academia; ranking systems and the gaming thereof; the pressure to generate grant money; civic engagement in large organisations; and ivory tower academia.

Article Six – BrokenBriar Affair Heating Up – Lawyers Involved

In this chapter we find out that the suspended students have hired a lawyer to fight their case. Lacking access to the professor's proprietary data sets, they argue they had no culpability for any research fraud. We also discuss the culture of doctoral studies in universities, and how these are often a poisoned pill for those seeking a career in academia. Ethical issues introduced here include fraud in data gathering and presentation; legal issues in workplace dynamics; 'sacrificial lambs'; career potential and how it can be damaged; career stability; performance metrics; the differential between jobs available and graduates produced; and the plague of short-term contracts in higher education and beyond.

Article Seven – Senior University Members Implicated in Growing Scandal

In this chapter we find out that an anonymous hacker is accessing university systems and leaking the information contained within to the press. Included in these leaks are emails from senior members of the university who encouraged Professor Blackbriar to fudge his data results to secure research grant funding. Ethical issues introduced here include the extent to which data can be sanitised without compromising its integrity; fixed thresholds for 'significance'; the categorisation of outlier data and the mental biases that influence the same; disclaimers; the security of data online; pilot studies and pilot data; dissemination of confidential information online; and file sharing generally.

Article Eight – Drunken Professor Lashes Out in Twitter Storm

In this chapter we find Professor Blackbriar's Blether account filling up with allegations against co-workers, superiors, and his postgraduate students. He later claims the account was hacked. Further emails leaked by the university hacker reveal more pressure to publish from the principal.

Ethical issues introduced here include social media use; 'being hacked' to avoid the consequences of misconduct; pressure from above; the factionalism of social medial; algorithmic assessment of online opinion; fear as a motivation tool; publication ethics for co-authors; and the cherry-picking of publication data.

Article Nine – Culture of Fear and Nepotism at the University

Additional leaked emails reveal the culture of fear at the university and the personal profiles sent around about the applicants for Blackbriar's research programmes. We find the chair of the research ethics committee engaging in casual misogyny and warning of the financial implications of hiring someone with visual impairments. Ethical issues introduced here include freedom of speech through anonymity; social proofing; artificial intelligence and the 'resume sift'; the impact of influencers; social media trolling; Deep Fakes; censorship; the impact of league tables on graduate prospects; discrimination in recruitment; the culture of fear in organisations; nepotism; conflicts of interest; and demographic equality.

Article Ten – Witch-Hunts at the University – IT Crackdown Causes Criticisms

In this chapter we learn of how the university is attempting to silence dissent and plug leaks by restricting access to outside computer services. Students are suspended as they work around the restrictions, but the university remains unrepentant. Ethical issues introduced here include codes of conduct; access to information; proportionality in punishment; name and shame; plagiarism in the age of artificial intelligence; ethical AI policies; shared resources and the tragedy of the commons; retroactive changes in terms and agreements; and bypassing corporate restrictions.

Article Eleven – Hacker Is Postgraduate Student

The hacker is revealed to be one of the postgrad students implicated in the growing scandal. The use of social network surveillance and the extraction of metadata is discussed to explain how the hacker was caught. Issues of computer security are likewise discussed. Ethical issues introduced here include disclosure for 'the public good'; the extraction of metadata online; the illusion of anonymity; social engineering; system security; computer misuse legislation; verifiability of whistleblower information; monitoring of computing resources; EULAs and agreement; and informed consent.

Article Twelve – Clean-Out at Scandal-Linked Journals

The journals where Blackbriar and his students published their results have a clean-out, firing peer reviewers and deputy editors. Conflicts of interest in the peer review process are cited as justification, and the role of academic paywalls in the publication process is discussed with reference to both commercial and open access outlets. Ethical issues introduced here include the role of peer review; editorial assent; the cost of journal subscriptions and access to information; open access; pools of expertise; generative AI in journal output; gift cultures; the rejection economy; and the value of reputation.

Article Thirteen – Student Journalism Outs Senior Academics

Serious conflicts of interest within the industry collaborations of Professor Blackbriar are revealed. Student journalists on Facebook explore the Internet Archive and find committee membership lists and minutes that show a link between university academics and the companies that provided funds. Ethical issues introduced here include board membership for senior academics; freedom of information legislation; generalising pilot data; conflicts of interest; corporate funding of academic research; hiding information online; the deep internet; internet archival; and the persistence of resources.

Article Fourteen – Resignations All Around at the University of Dunglen

Many senior members of the university resign and move on to other posts. The consequences of the scandal at the university for remaining staff and students are discussed. Ethical issues here include use of honorifics; golden parachutes; hand-tying clauses such as no-compete and non-disparagement; PhD funding transferability; the idea of the 'bus factor'; fragility of data processing chains; recruitment and institutional reputation; staff churn and attrition; and lost institutional knowledge.

POSTSCRIPTS

Completely new to this book are several postscript articles that are set approximately five years after the scandal came to public attention. They do not come with any additional commentary from the authors and are provided purely for wider context. Primarily they are intended to surface some perspectives that may not have 'served the narrative' of the reporting. They do not change anything in the core story but still offer insights that the participants felt worth sharing, all those years later.

Postscript One – Sharon McAlpine Interview

This interview is a supporting document that gives the deeper perspective of one of the characters within the study. This is Sharon McAlpine, talking about her experience as a young mother in academia, and how the systems she worked within contributed both to her role in the scandal and her general feelings about working in research.

Postscript Two – Stan Templemore Interview

Stan Templemore was a background character within the study – talked about, but rarely talked to. This interview serves to give his perspective on the scandal. It gives the reader some insight into what bystanders had to contend with and the way in which the fortunes of the principal players overshadowed the damage to everyone else.

Postscript Three – Jack McKracken Interview

Jack McKracken was the author of each of the newspaper articles, and after he broke the story of the scandal, he found himself riding a significant upward career trajectory. His perspective was largely un-examined within the core case study – after all, why would he critique his own work when it was doing so much good for the paper. This interview is his chance to reflect on his own coverage with the benefit of a little distance.

Postscript Four – Professor Blackbriar Obituary

Professor Blackbriar did not live to see his obituary – few people, after all, do. He was, throughout the study, remarkably tight-lipped on the whole. Here we may reflect on his epitaph and what others may say of us when our own time comes. Would Blackbriar's obituary have been happier, or more laudatory, had he behaved differently in his work life? What might we all have time to change in our own professional practice, and is that change worth enacting?

CAST OF CHARACTERS

The following is the cast of characters for the case study. These are introduced throughout the course of the articles, but the full list is presented for reference. Some minor spoilers are present in the descriptions for certain characters, and as such it is not advised to present this list in its entirety when the case study is being discussed for the first time.

1. Jack McKracken – Reporter and presented author of each of these articles.

2. Professor John Blackbriar – Academic at the centre of the crisis.
3. Professor Sir David Tumblewood – Principal and Vice Chancellor of the University of Dunglen.
4. Sharon McAlpine – PhD Student and Peer Reviewer.
5. James Duncan – PhD Student, Peer Reviewer, and Hacker.
6. Professor Callum Sunderland – Research ethics expert at the University of Alba.
7. Chuma Hassan – Spokesperson for the European Funding Council.
8. Stan Templemore – Budding PhD candidate and current masters student.
9. Vanessa Haynes – Research Director at the North Sea Algorithmic Exploration project.
10. Emeritus Professor Joanne Clement – Professor in the Anthropology of Volunteering.
11. Professor Elizabeth Burke – Former Principal of the University of Dunglen.
12. Professor Derek Taylor – Head of the Dunglen Business School.
13. Karan Chandra – Lawyer to the PhD students.
14. Dr. Jake Dymock – Employment Consultant to the University of Alba.
15. Professor Ian McManus – Head of the Research Ethics Board for the University of Dunglen.
16. Sir Gideon Lazenby – University of Dunglen Treasurer.
17. Terry Holmes – Head of IT at the university.
18. Detective Inspector Cameron – Police Scotland office.
19. Dr. Paula McCrane – Deputy Editor of the Journal of Deep Sea Oil Exploration.
20. Dr. David Sumner – Editor in Chief of the Journal of Deep Sea Oil Exploration.
21. Derek Simmons – Vice President of Industry Collaboration for ScotOil.

22. Professor Helen Hackett – Replacement Vice Chancellor of the University of Dunglen.

The cast of characters and the map to the scandal will hopefully help readers navigate these in the form intended.

We also avoid the use of any real universities or people in this case study. Where you see references to Blether, think of it like a Scottish version of Twitter. Other real-life social networks and companies have been renamed to fictional composite alternatives. This holds true only for the fictionalised aspects of this work – everything else will anchor into real-life examples that are supported with documentary evidence in the bibliography.

REFERENCES

Epstein, R. G. (1994a). The case of the killer robot (part 1). *ACM SIGCAS Computers and Society, 24*(3), 20–28.

Epstein, R. G. (1994b). The case of the killer robot (part 2). *ACM SIGCAS Computers and Society, 24*(4), 12–32.

Heron, M. J., & Belford, P. H. (2015). A practitioner reflection on teaching computer ethics with case studies and psychology. *Brookes eJournal of Learning and Teaching, 7*(1).

CHAPTER 2

Using the Case Study

This case study covers many important topics relating to computing and professional ethics. It can be used as is, with no reference to the commentary and within a classroom structure of bespoke design. Here though we present an overview as to how we have used it ourselves within classroom settings. We present a thematic overview along with suggestions for teaching exercises. This chapter has two main aims which we need to address before plunging you headfirst into the case study itself. The first of these aims is to highlight some of the important topics which we encounter, in context, throughout the case study. The second aim is to give examples of how the case study can be explored and debated during class sessions.

However, importantly here we advise you as an educator to allow the conversation to go wherever it goes. What we outline here is not the 'correct answer' to how people should interpret the study, but rather one set of common insights and perspectives that emerged as a result of our own interests and teaching experiences. Different people will identify different topics of note and will assign a different strength of moral judgement to the activities documented. The case study is intended as a prompt for discussion, but the book is not a list of directions as to finding topics worthy of that discussion.

This is a case study that is about computer ethics. Many of the issues we discuss are within an academic frame – this is both tractable to students and sufficiently transferable that it can be related to almost any knowledge-based scenario. Primarily though this book concerns itself with how technology impacts on our lives and the lives of anyone in a professional workplace. However, a work of this nature is less about the computers, and

DOI: 10.1201/9781003426172-2

more about the people that use computers. Ethical and moral responsibility for technology, how we talk about it and how we use it, is always to be found in the human component of the equation. While much of what we discuss is specifically within the professional framing of a university, the issues are intended to be largely universal. As part of the discussions to follow in the classroom, students – and teachers – should look for commonalities between their life experiences and those of the participants in the study. Relatively few people have published an academic paper, but they'll certainly have written an essay for an assignment or uploaded a YouTube video. The specifics as they exist in the case study are only examples of the general case. While most will not have experience of a domineering academic supervisor, many of us can think about a micromanaging boss and extrapolate.

With that in mind, we present this as a formal platform of study for the Scandal in Academia.

TOPICS AND ISSUES RAISED IN THE CASE STUDY

We reiterate that list of topics is not exhaustive. We have though divided the case study into four major themes to show a way in which the topics discussed can be effectively contextualised.

Data

- Data cleansing and data sanitisation
- Data fraud
- Online data security
- Cherry-picking publication data
- Closed data sets and access to information
- Dissemination of confidential information and file sharing online
- Fragility of data processing chains
- FOI legislation
- EULAs and agreement
- Reproducibility in science
- The impact of mental biases on decision-making

Culture

- Informed consent
- Legal issues in workplace dynamics
- Ethical and professional codes of conduct
- Reputational damage
- Power differentials in workplace relations
- Workplace bullying
- Discrimination in recruitment
- Culture of fear in organisations
- Nepotism
- Conflicts of interest
- Performance standards
- Career (in)stability and career potential
- Fear as a motivation tool
- Golden parachutes and hand-tying clauses
- Recruitment and institutional reputation
- Staff churn and attrition
- Lost institutional knowledge
- Value of reputation
- Demographic equality
- Freedom of speech through anonymity
- Impact of university league tables on graduate prospects
- Student evaluations

Knowledge

- Corporate interests in academia
- Impact of academic research on industry

- The role of postgraduate students in academia
- Scientific misconduct
- Publish or perish in academia
- Implications of funding model for academic research
- Peer review and editorial assent
- Cost of access to information – subscription and open access journals
- Gift culture of academia
- Publication ethics for co-authors
- Civic engagement
- Impact of ranking systems in academia
- Ivory tower academia
- Board memberships for senior academics
- Use of artificial intelligence
- Plagiarism

Technology

- Social media usage and impact on reputation of self and employer
- Accessibility issues
- The illusion of anonymity online
- Social proofing
- Censorship
- Hiding information online
- The deep internet
- Internet archival
- The persistence of resources
- Monitoring of computing resources
- Social engineering

- System security
- Ethics of whistleblowing

ACTORS ON TRIAL: WORKING THROUGH THE CASE STUDY

One of the benefits of this case study is that it introduces many topics for discussion with a fixed, understandable context. Rather than trying to frame ethics within philosophical schools or through abstract thought exercises, we present this as a more holistic, embedded approach to the topic of computer ethics.

The case study can be used as the basis for an active exercise. We have had considerable success by emulating a courtroom where each of the actors within the story are placed on trial for their role in the affair. This is to ascertain to individual satisfaction their perceived level of culpability. This can be carried out over a period of several classes. The suggested timescale below assumes that each session is of at least two-hour duration. For shorter tutorial sessions, this could be broken down into 14 sessions – one per article. Articles which are not core to the themes and concepts that one may wish to discuss can be summarised or assigned as extra reading.

A SESSION-BY-SESSION ANALYSIS

Session One – Introduction

An approach with which the authors have had success is to release the newspaper articles on an incremental basis, at different points during the course. Each hour of classroom instruction can be broken up into a separate, self-contained structure. We recommend at least two hours per session to give room for students to fully immerse themselves into the discussion – that would give a session that covers two articles in the requisite depth.

The process for each hour can include the following set elements:

1. Hand out a newspaper article and give students time to read and make notes.
2. Discuss relevant ethical topics which have been introduced via the newspaper articles.
3. Get students to research, in groups, some of these topics.

4. Have a debate about where blame can be apportioned based on what we know so far.
5. Identify similar experiences that students, or people they know, have experienced in their own lives.
6. Get students to fill in a logbook detailing their findings and opinions relating to the topics discussed.
7. Discuss their current opinions and judgements of the various actors implicated in the scandal.

The following articles should be part of this session:

1. A Scandal in Academia
2. Student Suspensions at Scandal-Ridden University

These introduce the following participants:

- Professor John Blackbriar – a preeminent researcher with over 30 years of experience.
- Vice Chancellor Professor Sir David Tumblewood – the head of the university and a personal friend of Blackbriar.
- Professor Ian MacManus – Chair of the Research Ethics Committee at the university.
- Sharon McAlpine – a PhD student of Blackbriar, accused of tainting the research data.
- James Duncan – another PhD student of Blackbriar, who is also accused of inappropriate data sanitisation.
- Professor Callum Sutherland – a senior academic from a rival university.
- Jack McKracken – the journalist responsible for writing the articles.

They also introduce the following institutions:

- The University of Dunglen – a post-1992 university fighting to establish itself as a serious research institution.

- ScotOil – an oil sector company which has funded much of Blackbriar's research.
- Dunglen Chronicle – the newspaper which has broken the story of alleged misconduct at the University of Dunglen.

These articles introduce the following broad topics which can form the basis of discussion:

- Professional and Ethical Codes of Conduct (e.g. BCS and ACM)
- Corporate interests in academia
- Reputational damage
- Issue of reproducibility in science
- Data sanitisation and data cleansing

At this point, we don't know much about what has actually happened, who has done what, and whether or not there has actually been any wrongdoing at all. However, opinions have probably already started to form. This is a good point at which to introduce the main elements of the BCS and/or ACM codes of conduct, and to discuss professional and ethical codes of conduct in general. How clear do the students think the codes are? How realistic is it to expect employees or society members to adhere to all aspects of the codes at all times? Can they think of or find any examples of individuals who have fallen foul of these standards? Have they found any cases where there are exonerating circumstances such that they think the person was correct to break the code? What about their own institutional contexts – how much do they know of the ethical codes under which they are already bound?

We can also ask students to assign their 'first impressions' of guilt or culpability, on the basis of what we have heard so far. They should also provide reasons. Students can then be divided into groups, to research the topics encountered in order to prepare their cases in defence of, or prosecution of, the following actors:

- Professor Blackbriar
- Sharon McAlpine

- James Duncan
- ScotOil

As we know very little about the case yet, and particularly about the two PhD students, they could be taken together at this point in the case study.

Topics which students could research to inform the debate include:

- What is data sanitisation?
- What counts as acceptable data sanitisation/cleansing?
- At what point does data manipulation count as fraud?
- What are proprietary data sets?
- What ethical issues are associated with making data sets proprietary?
- What is the typical contract/relationship between a PhD student and their primary supervisor?
- Who holds ultimate responsibility for their research findings and publication output?
- What is the power relationship between a PhD student and a professor?
- How might power dynamics impact on the behaviour in a work place?
- Are power imbalances potentially exonerating circumstances?
- In what ways might there be a conflict of interests between the research community and the corporation(s) who fund their research, and why?
- At present, we do not know much. However, the reporting of it has already caused reputational damage to at least three individuals. What are the ethics of a newspaper creating this kind of early impression?

Each group should present their case in defence or prosecution. After each group has presented, a vote should be taken regarding who is believed to be most culpable (if anyone) at this point in the case study. Record the numbers if possible – it can be instructive to view the change in attitudes as the scenario is explored.

It is also recommended that students are asked to keep a logbook. They can use this to record the main arguments made, the current view of the class, and their own opinions regarding where they believe guilt lies at this point, and why. This should be added to incrementally after each newspaper article is introduced. As an 'extra exercise', students should reflect on the revelations from the week and make a note in their logbooks as to what the participants could have done, if anything, to have prevented the scandal taking the shape it did. What could have been done better, what could have gone worse, and what do they think they might have done under the same circumstances?

Session Two

The following two articles are most appropriate for this session:

1. Multimillion-Pound Consequences for Research Fiddle
2. Students Speak Out

These articles introduce the following new characters:

- Chuma Hassan – spokesperson for the European Funding Council (EFC)
- Vanessa Haynes – Research Director of the North Sea Algorithmic Exploration (NSAE) project

These articles include the following topics which can form the basis of discussion or argument building for a debate:

- Corporate influence in academia
- Impact of academia on industry
- Speculative investment on the basis of uncorroborated research
- Mental biases which impact on decision-making
- Workplace relations
- Power differentials

- The role of postgraduate students in modern academia
- Closed data sets
- The publication imperative
- Workplace bullying
- Issues of accessibility of information resources

Since Session One, we have learned more about the complex relationship between academia and industry. We have also learned more about several aspects of workplace relations, and about some issues relating to accessibility and closed data sets.

This is a good time to have a deeper discussion about workplace culture. During the first session, there was some consideration of the professional relationship between PhD candidates and their supervisors. Now we have heard in much greater detail – from the students' perspectives – about the specific relationship between Professor Blackbriar and the students he supervises. Does this sound like a good working environment for PhD students? Does this sound like a good working environment for professors? What factors might be contributing to this workplace culture? Can students identify any external factors which impact on the workplace culture and pressures? How have their earlier impressions been changed on the basis of new information? Why did they change, or why didn't they change? What do they want to see as evidence that they don't already have?

After this initial discussion a new debate can be conducted, taking into account new revelations. Students can be divided into groups to create cases regarding what responsibility (if any) should be ascribed to the following institutions and actors. Each group should focus on one of the following, with a view to acting as either the prosecution or the defence:

- Professor Blackbriar
- Sharon McAlpine
- John Duncan
- Vice Chancellor Professor Sir David Tumblewood
- ScotOil

Topics which students could research to inform the debate could include:

- Research funding models – public and private
- Corporate funding of research and impact on dissemination and reproducibility of results
- Career paths and career stability in academia
- PhD funding models
- Desirability of funded PhD positions
- The supply vs demand of PhDs in various subject areas.
- Contractual publishing requirements and publishing 'norms' in academia
- Research grant funding – application processes and success rates
- Research grant funding allocation – how universities actually distribute research funding
- Legislation relating to employment rights and policies for handling workplace bullying

After each group has presented in prosecution or defence, a vote should be taken regarding who is believed to be most culpable (if anyone) at this point in the case study. Compare and contrast this to the results from the previous session – differences can be profitably discussed here, and on an ongoing basis.

It is also recommended that students are asked to update their logbook. They can use this to record the main arguments made, the current view of the class, and their own opinions regarding where they believe guilt lies at this point, and why. They should also record how their views have changed in any way based on the additional information they have received this session. As before, they should reflect on what they would have done in the same circumstances, and who could have done what to improve the situation as it is revealed.

Session Three

The following two articles are most appropriate for this session:

1. Leaked Minute Lays Bare University Culture
2. BrokenBriar Affair Heating Up – Lawyers Involved

The following new characters are introduced:

- Emeritus Professor Joanne Clement – former employee of the university, previously an anthropology professor
- Professor Derek Taylor – Head of the Dunglen Business School at the university
- Karan Chandra – a local lawyer representing the suspended PhD students
- Dr. Jake Dymock – an employment consultant at the University of Alba

These articles include the following topics which can form the basis of discussion or argument building for a debate:

- The ethics of whistleblowing
- Impact of quantification in academia
- Ranking systems and their susceptibility to gaming
- Pressure to generate grant funding
- Fixed performance standards
- Civic engagement in large organisations
- Ivory tower academia
- Fraud in data gathering and presentation
- Legal issues in workplace dynamics
- Sacrificial lambs
- Career potential and reputational damage
- Oversupply of PhDs
- Short-term contracts and their impact

Thanks to the two new articles, we have more details regarding the university culture, job (in)security, and the various pressures on staff and postgraduate students. The discussion about workplace culture from the previous session can be revisited in this session. What do students think that the information from these new articles adds to the understanding of the workplace culture? What do they think of the reliability of the sources of the information? Have the new revelations altered their views on any of the actors concerned? What are their opinions on the reliability and ethics of whistleblowing?

After this initial discussion, a new debate can be conducted, on the basis of what is known so far. Students can be divided into groups to create cases regarding what responsibility (if any) should be ascribed to the following institutions and actors. Each group should focus on one of the following, as either prosecution or defence:

- Professor Blackbriar
- Sharon McAlpine
- John Duncan
- Vice Chancellor Professor Sir David Tumblewood
- ScotOil

Topics which students could research to inform the debate could include:

- Further research into some of the suggested topics from the previous session:
  - Research funding models – public and private
  - Career paths and career stability in academia
  - PhD funding models, desirability of funded PhD positions, and the supply vs demand of PhDs in various subject areas.
- The Research Output Evaluation Framework (in real life, known as the Research Evaluation Framework, or REF) and its impact on the types of research which is supported and prioritised.
- Ethical and societal implications of how research focus is impacted by both funding requirements and corporate priorities.

After each group has presented in prosecution or defence, a vote should be taken regarding who is believed to be most culpable (if anyone) at this point in the case study. Again, compare and contrast the view in this session against that of previous sessions.

It is also recommended that students are asked to update their logbook. As before, they should focus on recording the evolution of opinions, thinking about how they would have behaved under similar pressures, and contemplating what could have been done better by everyone involved.

Session Four

The following two articles are most appropriate for this session:

1. Senior University Members Implicated in Growing Scandal
2. Drunken Professor Lashes Out

These articles introduce some new characters:

- Sir Gideon Lazenby – Treasurer at the University of Dunglen
- Nexus Energy Group –an oil company with extensive interest in the North Sea
- Nemesis – an anonymous hacker who is leaking private emails to the press

These articles include the following topics which can form the basis of discussion or argument building for a debate:

- The ethics of whistleblowing
- Pressure to publish
- Corporate impact on academic research
- Reproducibility of research results
- Memes
- Deep fakes

From these two new articles we have discovered – assuming the leaked emails are genuine – that the Vice Chancellor and the Chair of the Ethics Committee were both aware of Blackbriar's concerns regarding his data. Not only were they aware, but they pressured him to publish the results, citing the threat of losing commercial funding for future research. The discussion about workplace culture can again be revisited in light of these new revelations. Does anything in the emails explain or exonerate Blackbriar's actions and behaviour in any way?

This is also a good point at which to have a discussion about social media culture and the problematic aspects of this. At least two individuals implicated in this scandal have become the subject of memes and/or deep fakes. What do the students think of meme culture? What do they think of deep fakes? Is the creation and deployment of either of these ethical? What are the implications in terms of reputational damage? Does the supposed anonymity of the Internet have an impact on the debate? The previous discussion regarding the ethics of whistleblowing can also be revisited in light of the new leaks.

After this initial discussion, a new debate can be conducted, on the basis of what is known so far. As usual, students can be divided into groups to create cases regarding what responsibility (if any) should be ascribed to the following institutions and actors. Each group should focus on one of the following:

- Professor Blackbriar
- Sharon McAlpine
- John Duncan
- Vice Chancellor Professor Sir David Tumblewood
- ScotOil
- Professor Ian MacManus
- Sir Gideon Lazenby

Topics which students could research to inform the debate could include:

- Further research into some of the suggested topics from previous sessions:
    - Corporate funding of research and impact on dissemination and reproducibility of results.
    - The ethics of whistleblowing.
- Corporate funding of academic research and pressure to find suitable results. Are there any controversial cases relating to this?

- Social media arguments and the impact on careers and reputations.
- The ethics of cancel culture.
- The ethics of creating deep fakes.
- The impact of deep fakes on online debates and levels of trust in society.

In what is now the normal fashion, a vote should be taken regarding who is believed to be most culpable (if anyone) at this point in the case study, and the results compared against previous sessions. Students should then update their logbooks.

Session Five

The following two articles are most appropriate for this session:

1. Senior Culture of Fear and Nepotism at the University
2. Witch-Hunts at the University – IT Crackdown Causes Criticisms

The following new characters are introduced:

- Dirk Tumblewood – unsuccessful PhD applicant and Sir David Tumblewood's nephew
- Stan Templemore – unsuccessful PhD applicant and current Data Mining student
- Terry Holmes – Head of Information Technology at the University of Dunglen

These articles include the following topics which can form the basis of discussion or argument building for a debate:

- Employment discrimination and recruitment biases
- Nepotism
- Use of AI tools in recruitment
- Acceptable Use Policies
- Large language models

From these new articles, we are now aware of a culture of growing fear and suspicion at the university. The hacks and leaks have resulted in changes to the IT Acceptable Usage Policy, and restricted the internet usage of all students (whether staff are also affected is unclear). This includes access to several legal sources of entertainment. Students who attempt to bypass these restrictions, which many feel are unfair, are disciplined and threatened with further sanctions. They are expected to report on any friends who are also guilty of breaking the new AUP. We are also aware – with the caveat that the emails released by the hacker are genuine – of the views of the Head of the Research Ethics Committee regarding the shortlisted PhD candidates – views which seem to be based not wholly on the relevant factors concerning their suitability for the course of study.

Artificial intelligence is mentioned in relation to CV sifting and also in relation to both student plagiarism and legitimate student research. It is worth broadening discussion to include discrimination in hiring practices and the protections provided in law. This could include a discussion of the (unintended) biases which can be built into AI which often discriminate against applicants before a human becomes involved in the process.

What do the students think of the pressure applied to hire the Vice Chancellor's nephew? What do they think of the comments made relating to Sharon MacAlpine? Particularly those regarding her appearance? What about the potential of her family to delay the research by placing demands on her time? Is it reasonable to take a potential future delay due to hypothetical maternity leave into account? What are their thoughts on the comments relating to the inconveniences and expense of hiring James Duncan? Has the university been fair in clamping down on internet use due to the leaks and current scrutiny? What are the ethics of using large language models?

How would they feel if they knew their own professors were talking about them in this manner?

After the discussion, we follow the standard template of forming groups. This time, focusing on the following actors:

- Professor Blackbriar
- Sharon McAlpine
- John Duncan

- Vice Chancellor Professor Sir David Tumblewood
- ScotOil
- Professor Ian MacManus
- Sir Gideon Lazenby

Topics which students could research to inform the debate could include:

- Employment discrimination legislation.
- Disability discrimination legislation.
- Legislation relating to the misuse of computers.
- The relationship between universities and their students.
- What services do students have the right to expect from their university?
- Is it acceptable to have conditions changed partway through an academic year?
- Is there any legislation which protects student access to the online resources which they require in order to conduct their studies?

This is followed by the vote, as usual, and the updating of student logbooks.

Session Six

The following two articles are most appropriate for this session:

1. Hacker is Postgraduate Student
2. Clean-Out at Scandal-Linked Journals

We have several new players, as usual:

- Dr Paula McCrane – Editor of International Journal of Extreme Extraction
- Dr David Summer – Editor of the Journal of Deep Sea Oil Exploration

These articles include the following topics which can form the basis of discussion or argument building for a debate:

- Social engineering
- Entrapment
- Social Media Usage Policies
- Computer Misuse Act
- User Consent
- Peer review process for journals
- Journal publishing models – paid vs open access
- Gift cultures in academia

Thanks to these new articles we now know that the hacker was James Duncan – one of the PhD students implicated in the scandal. Does this revelation alter any opinions in any way regarding his level of responsibility with regard to the alleged data fraud? What is the reasoning for this? In addition to discussing the unveiling of the whistleblower and the implications for the rest of the case, there are various other issues to discuss.

Social engineering is something with which some of the students may have personal experience. How can they guard against it? What responsibilities do we have to do our utmost to prevent ourselves from being easy prey against hackers and social engineering? Social Media Policies are becoming more common, with employers seeking to dictate what employees say about them online. Possibly also dictating what they can and cannot post about other topics even on their personal accounts. Is this fair? Is it compatible with the right to freedom of speech? How far does that right extend, and how far should it extend? What about freedom of speech versus freedom from harm?

Discussion of the academic publishing model and gift culture in academia opens interesting philosophical discussions about how research is funded and how the results are disseminated. Is it morally right that the results of publicly (or corporately) funded research are most often published behind a private paywall? Should all public research be made publicly available for the benefit of society in general? How often have students tried to access papers for their own work and found them inaccessible?

After the discussion, there is the usual debate. Here, we recommend it focus on one of the following:

- Professor Blackbriar
- Sharon McAlpine
- John Duncan
- Vice Chancellor Professor Sir David Tumblewood
- ScotOil
- Professor Ian MacManus
- Sir Gideon Lazenby

Topics which students could research to inform the debate could include:

- Social engineering
- Social engineering as entrapment
- Monitoring of personal social media use by companies
- Conflicts of interest
- After each group has presented, the usual vote should be conducted and logbooks updated.

Session Seven

In this final session, release these articles:

1. Student Journalism Outs Senior Academics
2. Resignations All Around at University of Dunglen

Our final few characters are introduced at this point.

- Derek Simmons – Vice President of Industry Collaboration at ScotOil
- Professor Hackett – a Professor at the University of Dunglen

These articles include the following topics which can form the basis of discussion or argument building for a debate:

- Conflicts of interest
- Workplace bullying
- The dark web
- Internet Archive and the permanence of the Internet
- Golden parachutes and non-disclosure type agreements

Thanks to some student investigations involving the internet archives, we have discovered further evidence of inappropriate conduct by senior academics. Sir Gideon Lazenby, who as the University Treasurer has an interest in the University receiving project funding, is also on the Board of Directors of ScotOil, and played a role in persuading ScotOil to fund the project. Professor McManus was one of the reviewers who approved the funding for the project, rather than recusing himself due to a conflict of interest. There is also further evidence of pressure being applied to Blackbriar to move forward with the algorithm despite uncertain results.

This is a good point to revisit the earlier discussions on workplace culture and bullying, and conflicts of interest. Have opinions towards any of those named in the scandal changed as a result of these new revelations? It is also a good time to discuss reputational damage and the permanence of online information. What are the ethics of trying to delete or hide previous posts or articles? What are the ethics of searching for information that individuals have tried to delete? Finally, it is also a good time to discuss the impact and fallout from the scandal on the individuals involved. Clearly, the senior academics implicated have either found sideways moves or have at least been cushioned from the financial impact of their change in position thanks to golden parachutes. The PhD candidates, however, appear to have received no recompense. Is this fair?

The task at the final stage here is to begin moving towards a judgement as to culpability. For the first article discussion in this session, conduct the debate as normal focusing on the following actors.

- Professor Blackbriar
- Sharon McAlpine

- John Duncan
- Vice Chancellor Professor Sir David Tumblewood
- ScotOil
- Professor Ian MacManus
- Sir Gideon Lazenby

For the last session, groups should focus on ranking participants into a top five of culpability, along with reasons for their ranking. They can choose anyone to be in this list, regardless of what prominence they may have had in the case study. They then present this list in groups, with each unique participant being added to a shared pool of responsibility constructed by the class. Finally, everyone should vote on who they think is most responsible from this selection, and whether the punishment – or otherwise – they received was proportional. There is no right answer here. We, as the authors, have our opinion. It couldn't matter less though what **we** think.

Now that we have a structure you might want to follow, we can finally get on to the case study!

CHAPTER 3

A Scandal in Academia

NEWSPAPER ARTICLE

An exclusive report by Jack McKracken for the Dunglen Chronicle

The city of Dunglen was rocked today when one of its most respected academics was suspended as part of an investigation into 'gross misconduct' and 'academic fraud'. The lecturer in question is Professor John Blackbriar, OBE, a world-renowned expert in the area of algorithm-based oil extraction. His work is used by companies all over the world to maximise the amount of raw crude that can be recovered from ageing fields in deep water environments.

Professor Blackbriar has been employed by the University of Dunglen for 30 years. He graduated from the University of Edinburgh in 1982 with a joint degree in mathematics and chemistry. He worked for various oil companies for ten years before joining the University of Dunglen in 1994 as a research associate. In 1999, he was awarded his PhD for his thesis 'Onwards and Upwards: Simulated Annealing as a Proxy for Deep Exploration'. He has published over 250 papers in world leading journals and is first author on 80 of these. According to the academic search engine Google Scholar, his papers have been cited over 41,500 times. He is responsible for generating 80 million pounds in research grants for the university, and his commercial links with the oil and gas sector have given the University of Dunglen an edge over competitors for many years.

The scandal has been brewing at the university for several weeks according to sources close to the vice chancellor. Professor Sir David Tumblewood's office today released a statement: 'Serious allegations of academic fraud have

DOI: 10.1201/9781003426172-3

been levelled at one of our most senior academics. We have no doubt as to his integrity and the quality of his work over the years. However, the policies and bylaws of the university are such that in situations like these we must suspend the academic in question while we perform a full investigation. We hope Professor Blackbriar will be exonerated at the conclusion of this procedure, but we cannot comment on specifics as to time-frame or process'. Talking personally to the Chronicle, Professor Sir David Tumblewood said 'John has been a good friend of mine for ten years, ever since I arrived at the University. I fully expect that our investigation will completely exonerate him. He's an honourable man, a dedicated scholar, and one of this university's key assets. However, we have to investigate all alleged breaches of this university's ethical code of conduct'.

News of the allegations have created rumbles elsewhere in the country – a senior source at ScotOil said to us 'We've been having troubles for a while in making his algorithms provide us with the same measures of confidence that his papers suggest. At the moment we are still putting that down to methodological differences. Still though, it makes you wonder'.

This source added, 'If the allegations are true, this might end up costing us hundreds of millions. It's a costly procedure to extend the life of oil-rigs on speculative exploration. We'd hoped that our investment would be rewarded by an increase in production, because that's what the Blackbriar Algorithm told us would happen. And to be honest, it's what we expected given his previous track record'.

Like most universities, academics at Dunglen are expected to adhere to a formal code of conduct that governs their behaviour in various circumstances. Professor Ian McManus, chair of the Research Ethics Committee at the University, explained this to the Chronicle.

'Everyone working at the university is a decent, honest professional. Everyone at the university can be trusted to do the right thing. However, the problem comes in when people don't agree on what the right thing is. Our own personal morality might incline us to one course of action, whereas the personal moral code of another may direct them down a different course. It's important within a large organization and society generally that there is a code of conduct that supersedes our own personal feelings. In society, we call those laws. In the university, we call it a code of conduct'.

As a professional academic working in the fields of engineering, computing, and chemistry, Professor Blackbriar is bound by several official codes of conduct. As a fellow of the British Computer Society, he is required to adhere to the BCS Code of Conduct. The Royal Academy of Engineering requires that members adhere to their Statement of Ethical Principles. The Royal Society of

Chemistry also requires a Code of Conduct from its members.

'Those represent professional ethical codes', continues McManus, 'And membership of those societies is predicated on adherence. The University itself is exactly the same – to be a member of our research community you must follow our Ethical Code of Conduct. This covers everything from how experiments involving human participants must be conducted to how authorship of publications should be assigned. It deals with how sensitive data must be protected, and how individuals must conduct themselves when publicizing their work. It's a comprehensive policy. If these allegations against John are true, and I sincerely hope that they are not, then he would be in violation of that code of conduct. Until we can determine whether or not that's the case, he's been suspended with full pay'.

When asked whether or not the professional organisations to which Blackbriar belongs would be instituting similar inquiries, McManus said, 'I can't comment on that – but we all expect the highest ethical behaviour on behalf of our representatives, colleagues and members. I would be surprised if they didn't take an active interest in what's going on here. I can't speak to whether or not he has violated any of their rules, but if he has then his membership of those organization is liable to be suspended too'.

The work in question was initially supported by the European Funding Council (EFC) as part of a decade-long monetary package worth around 50 million euros. Post Brexit, this has been supplemented by funding from industry partners. The EFC retains a stake in the project and its ongoing success. It is alleged that the misconduct at the university is significant enough to have 'Europe-wide implications', but solid details remain elusive.

Professor John Blackbriar was last night unavailable for comment.

ETHICS, MORALS, AND LAWS

Let's begin our exploration of this case study by reaffirming our philosophy – you won't agree with everything that we say in this book. That's good – it's healthy, and readers should not fight their instinct to pick fault with the arguments put forward here. This is not a case study concerned with right versus wrong, or good versus evil. It's about the **context** of ethics, especially those that touch on the design and deployment of computers and their software. By design, it's full of grey areas – no significant person escapes the scenario without having at least some role in the emerging crisis. It's also not about finding the right person to blame for what happened – everyone is to blame, in their own way. Some are more to blame than others, but who falls into that category should be a matter of vigorous discussion and disagreement. To paraphrase Groucho Marx, we have our views, and if you don't like them – well, we have others.

We should also start off with a working definition of some terms. We're not going to spend a lot of time on the philosophy of morality or the way in which ethical codes are constructed. It's not relevant to the way we approach the topic. Even if it were it's unlikely to be very interesting to those who just want to have a discussion about the implications of the actions of those in the case study. We do though have very particular ways in which we use the words ethics and morality.

These terms are often used interchangeably, but in this book they are two different things. We define morality as an emergent property that is the natural consequence of a thousand small experiences. For most people, the construction of our moral codes begins with the family – our parents instil certain values in us as a natural consequence of osmosis. When they tut at the television, point to a newspaper article, or adopt a particular tone when discussing an issue, it all rubs off on us. Children naturally want to please their parents, and so early on we learn the 'right' way to talk about particular kinds of issues.

As we grow a little, we enter into a bigger world where our parents aren't the be all and end all of our moral instruction. We meet other people who come from other families, and we find out that they don't think quite the same way as we do. We encounter social structures that encapsulate certain moral messages, and we find ourselves sometimes in alignment and sometimes in disagreement. Cognitive dissonance, the process by which two competing elements vie for dominance in our minds, comes into play and something has to give to bring our thoughts and actions into alignment. This is a reasonably well-supported psychological principle – as a rule, our minds will exert pressures on us to bring harmony between our thoughts and our actions. When this harmony is not achieved, we enter an unpleasant mental state known as dissonance (Festinger, 1957). The state also applies when we must decide between two contradictory beliefs. When dissonance is encountered, we have three options:

1. Reduce the importance attached to particular beliefs, so that the dissonant beliefs can be rejected as situational or irrelevant in context.
2. Add more consistent beliefs that outweigh the dissonant beliefs, so that the dissonant beliefs can be discarded as an unusual exception.
3. Change or reject the dissonant beliefs.

These are powerful forces – an important paper on the subject by Mahaffy (1996) discusses the power of the effect to change the way lesbian Christians dealt with the perceived dissonance between their religious belief and sexual orientation. Coping strategies to deal with the issue included recasting core beliefs or abandoning their religion entirely. Studies around homosexual evangelicals (Thumma, 1991) likewise show the power that dissonance has to internally deconstruct and reconstruct our sense of identity. Every day, as our views are being formed by comparison with those around us, we must resolve many incongruities until we arrive at our sense of what is right. That's our sense of morality.

The societal structures that impact on us can be small – at the level of a local community, or huge – the level of governments and religions. We have to navigate ourselves through a maze of these, picking up our sense of right and wrong as we do. This has the consequence that everyone's morality is different – we may have large degrees of commonality as a result of shared experience, but we've all led different lives and had different moral messages imprinted upon us – each with a different emphasis. The concept of moral relativism comes into play here. This is an often misunderstood principle that simply argues that there is no way to objectively agree on what is good and what is evil because these are subjective judgments and not drawn from any objective criteria.

One may filter down the universe to the smallest atoms and yet never encounter anything that deals with morality. Lacking a universal 'yardstick' against which we can measure a moral code, all we can do is judge other moral codes from the relative position of our own. There is no 'universal truth' that runs through our moral codes – they are a combination of that which was socially useful and advantageous from an evolutionary perspective. Those who subscribe to a belief in an all-powerful creator may be able to argue that their sense of morality is truly objective, but unless all parties can agree as to which creator and what their rules are, it remains a subjective, relative judgment.

That's all fine and dandy, but it doesn't make for a very good basis upon which to build a society.

That's where the role of ethics comes in. We, in this book, define ethics as simply a defined code of conduct – a formal set of rules to which all individuals within a particular community agree to adhere. In the biggest case, we might call them 'laws'. We may not agree with particular laws, and some may be against our own personal sense of morality, but we live in a society where we've all agreed that if we break the law there will be punishment. The rules and prohibitions embedded in religions – at least in modern society – are examples of ethical codes. We may not necessarily agree with them going by our gut feeling on what is right or wrong, but we all agree to taking on a common position so that we have a way of deciding on the 'right' course of action. Here again the example of those suffering dissonance as a result of incompatible actions and religious codes comes into relevance. Sometimes the codes to which we agree to adhere overlap with our personal opinions, and that's great. Sometimes they come into conflict, and how we resolve them is an important part of what defines us as people. Sometimes there are real consequences to violating a code of ethics. Sometimes there aren't. In all cases, we weigh up the right and wrong, balance it on our sense of morality, and decide what has to be done.

In our case study, we're mostly not talking about morality, laws, or religions. What we're talking about are smaller codes of professional ethics. These are codes to which we agree to obey upon accepting a particular role in society. Many professions have some kind of collegiate body, and this collegiate body will put together a code of conduct for all members who choose to join. Sometimes membership of this body is a prerequisite to practising in the field. The Law Society of Scotland for example regulates solicitors in Scotland, and this professional body sets exams that must be passed for membership to be obtained. Passing the exams and obeying the ethical code of the society are both mandatory. If a solicitor is 'struck off' the register of this society, they cannot formally practise, and often may not describe themselves as solicitors any more. Section 25A of the Solicitors (Scotland) Act 1980 enshrines this in law. While this is a very specific example, you'll find parallels all over the world applied to all kinds of professions. Violations of the ethical codes of professional, regulatory bodies like these can thus have serious consequences.

In comparison, there is no requirement that 'computer professionals' be regulated by any of the professional societies for the industry. There is no requirement for accreditation by the British Computing Society (BCS) or the Association of Computer Machinery (ACM) or the Institute of Electrical and Electronics Engineers (IEEE) before someone can write code, admin a system, or put up a website. Membership of bodies such as this is still dependent on adhering to the codes and bylaws of the organisation, but practising as a professional in the area is not predicated on membership. These organisations, lacking the legal authority to regulate, instead offer additional services to their members – access to specialised resources, career opportunities, or the ability to describe themselves as members in good standing. Most employers put forward formal ethical requirements within a contract, and that contract will usually contain clauses that require you to adhere to any other official policies put forward by the employer. Often, this will include an additional code of conduct. Failing to adhere to these codes can result in termination of employment.

There are layers and layers of these codes to which we must adhere, and they don't always fall into perfect alignment with each other, or with what we may ourselves see as 'moral behaviour'. A lawyer who believes his or her client to be guilty of a heinous crime may feel morally justified in 'throwing the case'. Ethically and professionally, they are bound by a code of conduct that requires them to give as effective legal representation as they can. If we all always agreed on what is the 'right' thing to do, we wouldn't spend so much time arguing about it.

That's an important point to take away from this case study – argumentation is important, because it allows us all to explore the differences in our assumptions, our mindsets, and our understanding of how best to resolve ethical incompatibility when we encounter it.

THE CRIMES OF JOHN BLACKBRIAR

What we have within our first article, 'A Scandal in Academia', is a serious allegation of academic fraud. The details of what that might entail though are currently hidden from us. A dignified professor with a lifetime of accomplishments is suspended by his employer for allegedly violating the university's research code. The professor is clearly a man of some substance, judging by the way that important people within the university talk about him. His work is also obviously of considerable financial value if the anonymous insider at the ScotOil company is to be believed.

Within the Scottish education system, as with the system in other parts of the world, the fact that John Blackbriar is a professor is significant. As a title, it carries more weight than it does elsewhere because it represents the highest academic rank that one can achieve. It generally goes from Lecturer, to Senior Lecturer, to Reader, to Professor. Only the latter are permitted as a general rule to use the Professor honorific. Several UK universities though are now moving to the American model of assistant professor, associate professor, and full professor. While growing in popularity, these remain comparatively rare at the time of writing and so we can generally work on the assumption that this is the traditional Scottish titling system. John Blackbriar is a professor, which implies he has a high standing within his institution.

The comments made by the university tell us quite a bit about what may have happened – a research code of ethics is usually designed to ensure several things:

1. Accuracy in reporting of results.
2. Probity in the conducting of trials, especially those involving other people.
3. Respect for the confidentially of research data and intellectual property.

Professor McManus has explicitly indicated that this is a potential violation of the research code of ethics, although whatever is being alleged is likely to impact on the professional bodies to which the professor belongs. We can begin to untangle this situation by looking at each of the three criteria in turn.

Accuracy in Reporting

This is a key element when communicating research results, and a bedrock of the scientific process. The nature of peer review and academic publication is such that verification of research data can take years, if it's ever done at all. There are all kinds of ways in which this principle could be violated – for example faking data, cherry-picking results, or incompetence in analysis.

The first of these is comparatively rare because of the extremely high cost to those who might be tempted. It's not unknown, however – take for instance the case of the once rising Netherlands star of Diederik Stapel (Bhattacharjee, 2013; Levelt et al., 2012) who confessed to faking portions of his research output in at least 55 papers. In at least ten student dissertations that his postgraduate students submitted for their doctorates, there was evidence of his fabricated data being used. His work had been highly influential and caught the attention of media outlets for its significance. His papers showed that train stations that are not well tended bring out racist tendencies in people. He showed that eating meat made people selfish, and that meat-eaters were generally less social than vegetarians. Unfortunately, when suspicions began to grow about the way in which his results were reported, it turns out his data was less than robust.

For several papers, the data he had obtained turned out not to show the results he wanted. When faced with the choice of abandoning his hypotheses, publishing negative results, or redoing the experiments, he decided to take a short cut – to create the data sets himself. He consciously engineered the data so as to give a meaningful result that supported his arguments, but with not so large an effect as to raise suspicion. When he didn't get the results he wanted still, or when he though the effects were too significant, he went back to the data sets and tweaked, tweaked, tweaked. When there were studies done that would have involved human participants, he did them on himself and extrapolated the results. So as to avoid suspicion with collaborators, he would run shared studies and report back on those accurate figures. His misconduct went back at least ten years. If you see a trace of Stapel in the case study we're discussing, it's not accidental.

He however is only one of many – Yoshitaki Fuji (Normile, 2012) was found in 2012 to have fabricated data in at least 172 scientific papers about anaesthesiology, which was a record at the time for incidents of scientific misconduct by a single individual. In 2010, Milena Penkowa, a neuroscientist at the University of Copenhagen, was judged to have introduced deliberate aspects of fraud into at least 15 papers. In 2012 again, Dipak Das at the University of Connecticut Health in Farmington was found guilty by a review board of 145 counts of fabrication or falsification of data (Retraction Watch, 2012). As of 2023, 184 of Joachim Boldt's papers have been retracted due to identified incidents of forgery (Retraction Watch, 2023).

We could fill an entire chapter with incidents like this alone. If our Professor has been 'tweaking' his data, then he's unlikely to find himself lonely. The Retraction Watch blog keeps a keen eye on incidents where published papers must be retracted because of fraud or misinterpretation – it's well worth keeping an equally keen eye on their posts if you want to get a feel for how often this kind of thing happens. The extent to which it occurs in general is very hard to estimate – fraud by its very nature is a hidden variable. Fanelli (2009) however suggests that a touch under 2% of scientists admit to having 'fabricated, falsified or modified data or results at least once', with over a third admitting 'other questionable research practices'. When asked about colleagues, the numbers rise considerably – to 14.12% for outright falsification, and 72% for questionable research practices.

Fanelli remarks in the abstract to her fascinating paper:

> Considering that these surveys ask sensitive questions and have other limitations, it appears likely that this is a conservative estimate of the true prevalence of scientific misconduct.

Outright fraud might be in the minority. Cherry-picking data, unfortunately, is far more widespread. Partially this is a result of the publication bias against negative results – many journals won't publish these at all, and papers that report on negative results rarely achieve the same kind of interest or prominence as the rest. This creates a natural incentive for people to simply not publish when the results are disappointing. This in turn creates a sense of false agreement in a field and slows down future research as investigative dead-ends are continually retried and re-explored. As such, researchers are sometimes tempted to search for significance even in insignificant data.

Cherry picking is insidious and revolves around selecting only the desirable data and discarding the awkward 'outliers' before an analysis is performed. By far, more positive results are published than negative results. Sterling (1959) found out that 97% of the studies in four major psychology journals had shown positive results at a statistically significant level – only 3% of the studies reported on negative results. Sterling later repeated the study and found that on the whole nothing had changed. Fanelli (2012) has shown that the problem is getting worse. Recognising significant differences between countries and disciplines, on average the frequency of the publication of positive results increased 22% in the period between 1990 and 2007. There is no evidence that the situation has improved over the intervening years, although calls for more acceptance of negativity have been growing (Echevarrıa et al., 2021). The positive publication bias isn't a result of people getting better at doing science, but instead a consequence of scientific publishing exhibiting its institutional biases.

Malpractice is difficult to detect unless other researchers have access to both the raw data and the criteria by which data was included for analysis. Even if all of this is available, there is an additional bias in the scientific process that judges replicating of previous results to be less 'valuable' than new and unique contributions to the literature. Unless a result is especially striking or surprising, it's often the case that it's never replicated at all. Some fields are better at reproducing than others but as a whole reproducibility is something more honoured in the breach than in the observance.

Incompetence in analysis too is a widespread problem – statistics are notoriously open to interpretation, and many researchers have little formal training in applying them to the analysis of data. This results in many papers being published where the statistical analysis is incorrect. It also means that the peer reviewers involved often lack the critical skills to identify the methodological issues before publication. The result is a confusing mess of studies and papers, some of which say one thing and others which say different. Altman (1994) wrote:

> When I tell friends outside medicine that many papers published in medical journals are misleading because of methodological weaknesses they are rightly shocked. Huge sums of money are spent annually on research that is seriously flawed through the use of inappropriate designs, unrepresentative samples, small samples, incorrect methods of analysis, and faulty interpretation. Errors are so varied that a whole paper on the topic, valuable as it is, is not comprehensive; in any case, many of those who make the errors are unlikely to read it.

This is a common story (Altman, 2002; Harris et al., 2009, 2011; Hopewell et al., 2010; Smith, 2014) – confined to one particular field, but one that has resonance elsewhere. The inclusion of bio-statisticians in peer review has improved the situation somewhat (Teixeira-Pinto, 2021), but most fields don't routinely employ this kind of approach.

Even at the level of examining the extent of fraud and error in cases as significant as retraction, there is little agreement. Nath et al. (2006) examined 395 retracted articles between 1982 and 2002. Of these, over 61% were retracted because of unintentional errors in analysis, sampling, and methodology, with a comparatively meagre 27.1% retracted because of outright scientific fraud. However, Fang et al. (2012) arrive at the opposite conclusion from the 2047 retracted papers that they examined: only 21.3% were attributable to error, with 43.4% being linked to fraud, suspected, or otherwise. The problem seems to be getting worse – Rapani et al. (2020) suggest that retractions in dental literature increased by almost 50% during the period of 2014–2018. Campos-Varela and Ruano-Ravina (2019) suggest a baseline retraction rate of one per 4000 papers, with over 1082 being retracted from PubMed within the period of 2013–2016. During the height of research into Coronavirus, papers were being retracted at a rate of one per 2500 (Yeo-Teh & Tang, 2021). The numbers may not seem horrendous, but bear in mind there's a lot of research out there and retracted papers continue to be cited long after they have been withdrawn (Bolland et al., 2022). Suffice to say – there's a lot of badly conducted science out there, and the much misunderstood principle of peer review does little in guarding against it (Smith, 2010). We may not know yet what Professor Blackbriar has done, but we know that there's a wide range of candidate sins that are common to the field.

Probity in the Conducting of Trials

The conducting of trials, especially those involving human participants, is a constant issue of ethical reflection in science. Medical research particularly involves much analysis and oversight to minimise the likely negative impacts for participants and is perhaps the most highly developed area of research ethics. It wasn't always so – infamous examples such as the Tuskegee Syphilis Experiments (Brandt, 1978) of the early 1930s show that concern for research participants wasn't always enshrined in trials. Similarly, psychology experiments such as the Milgram Experiment (Milgram, 1965) and the Stanford Prison Experiment (Zimbardo, 2011; Zimbardo et al., 1972) show a disregard for honesty and consequence for participants that is shocking to modern sensibilities.

Deceit about to what people were agreeing when they signed up for a research project was once commonplace. Many of the landmark experiments of social psychology would be unlikely to receive assent under the research ethics regimes of most universities now. A standard protocol is to require that all participants are fully briefed as to the study's intention at its beginning, and fully debriefed at its conclusion as to what their participation meant. Likewise, standard operating protocol for most human research these days requires the researcher to explicitly outline that participation is voluntary and consent may be withdrawn by the participant at any time during the trial with no negative consequence.

However, there are other aspects here beyond simply being honest and upfront with those volunteering – a reliable methodology for conducting a trial is mandatory, especially if it's involving multiple researchers, participants, and partners. There needs to be a consistent way that data is gathered and logged, and sometimes that can be difficult to do when there's any element of subjective assessment. Consider for example the standard 'pain chart' used by clinicians to determine just how much pain an individual may be in. It consists of a series of faces, each demonstrating a degree of pain, and invites the patient to show where on the chart they think the face most closely matches their own experience.

Assume a simple study where the intention is to place a particular condition somewhere on this chart by noting the average response of people suffering from it. Over a large enough group of people, it can be hoped that the answers given 'smooth out'. If there's an average of 7, which is gained by asking over 1000 people, then by and large that can be trusted to have roughly the same number of people over-estimating as there are under-estimating. The more data there is, the less outliers impact upon it.

Such things need to be carefully managed so that 'wishful thinking' doesn't make its way into the data. The raw data is precious – it has to be as clean and as accurate as possible to ensure that any further analysis is done on a solid foundation. We need to make sure people are asked the same questions, in the same order. We need to ensure the way in which they provide their answers is consistent from day to day, researcher to researcher, participant to participant. Even something as innocuous as changing the wording of a verbal question can influence the results.

Data contamination is also a serious issue in research. Researchers are people too, and they make mistakes. Sometimes they forget to record a value they're supposed to. Sometimes they just didn't capture a measurement and now it's gone. Sometimes data gets recorded incorrectly, or transcribed inaccurately. A good methodology for a study will have processes in place to calibrate the data recorded, and for dealing with identified flaws. Usually, they're removed from analysis, although in other circumstances they may be resampled or 'scaled'. Different studies will demand a different response to this, and it's important that all researchers are upfront about how data contamination is handled. Similarly, when data has been excluded from analysis, and how contamination was detected. This has to be transparent both in the design of the study and in any subsequent reporting of the work through publications and public outreach.

Respect for the Confidentiality of Research Data and Intellectual Property

Data, especially that coming from human beings, can be sensitive. Research participants provide their input into studies on the understanding that the data they generate will be treated with the appropriate level of confidentiality. For medical and psychological data, this is especially important. A research code of ethics will document how data is to be treated, stored, and transmitted. It is normal that researchers directly involved in a study will have access, where appropriate, to the raw data but that all other individuals receive only anonymised extracts. When the research data does not involve other people, there may be other requirements. There may be issues of commercial confidentiality or intellectual property ownership. Data generated via studies has its own particular value, and the source of the funding for studies will impact heavily on how the data may be used.

Commercial interests in research are common, and a researcher may find that their hands are tied with regard to reporting results. This may be especially true in cases where the results show something other than what their funding partners might desire. In such cases, the code of ethics will normally outline what is to be done with regard to publication of results. Finally, there is an issue regarding academic authorship – whose name goes on the paper. This is a specialised kind of 'respect for intellectual property' and one that is of keen interest to researchers. Different fields have their own conventions on how it works, but the location and prominence a name receives on a paper is significant. 'First Author Status' is important in many fields – it indicates that the first author named on a paper is the one to whom most of the authorial credit should be allocated. It matters too in tangible ways – being the main author on a paper is valuable for promotion, for credibility in funding applications, and in demonstrating 'research independence'.

The further down the 'pecking order' you are on the authorship list, the less of the 'credit' belongs to you. When your work is cited in other papers, many referencing styles serve to give prominence to the first named author. For example, using some intext citation systems, the first mention of a paper gets the full list of authors, such as (to use one of Dr. Heron's papers as an otherwise irrelevant example) in Heron, Hanson & Ricketts. (2013). In subsequent citations, this becomes Heron et al. (2013). The result is that the first author receives considerably more notice in cases of multiple citation than subsequent authors. First author status then is important in getting a researcher's name out there. Most research codes will have something about how authorship should be allocated, and there is a convention known as the Vancouver Recommendations that are generally appropriate. Journals to which papers are submitted will likely have further guidance.

We are left with a situation then where our professor has fallen foul of one or more of these issues. Being as how his work is about algorithmic extraction of crude oil, it's probably not an issue regarding human participation in his trials. This is 'computer science' and abstracted away from the level of flesh and blood. As a well-established professor, we might assume that he is unlikely to have fallen foul of any simple methodological issues with the work he has done – we know he's published 250 papers and first-authored 80 of them. By now he's almost certainly gotten the 'easy mistakes' out of his system. **Almost** certainly – as we saw above, stature is not necessarily a defence against this kind of methodological error.

Likewise incompetency in analysis – while such issues are rife, our professor has a joint degree in Mathematics and Chemistry. It's not impossible he's made enough mistakes in his publications to be worthy of investigation, but it seems unlikely. His work is highly algorithmic in nature which implies more objectiveness than subjectiveness – that's going to reduce the chance it's to do with 'wishful thinking' in gathering information. Also, the anonymous ScotOil representative has suggested his previous work has lived up to expectations.

What we're left with then is a small set of realistic possibilities:

1. Straight up academic fraud in terms of fabricating data or cherry-picking results.
2. Lack of respect for the confidentiality of research data.
3. Lack of respect for intellectual property.
4. Violation of authorship rights.

All of these are serious allegations. Why then would our professor risk his career and reputation? What is there to be gained from throwing away the future to participate in any of these shoddy deceptions?

Academia is a strange and unusual environment – one that makes little sense to outsiders in terms of the incentive and motivation structures. When Michael published his first couple of papers (Heron, 2012; Heron et al., 2013), he proudly exclaimed to his mother that he'd had then accepted. She asked, rather quizzically, 'Do they pay you for doing that?' and seemed somewhat bemused when he'd say 'No'. Much of what drives an academic onwards is not direct financial reward for the work they do, but is instead 'expectation of opportunities'. Publishing papers makes them available for other researchers to read and cite. Citations add to an academic's reputation, and a growing reputation in turn opens up future possibilities in terms of generating research funding or gaining employment in larger or 'better' institutions. Promotion too is based on the perceived reputation of the academic as well as the likelihood they can parlay that reputation into opportunities for the institution. Hence, the power of the first author position.

See how in our case study that Blackbriar's commercial contacts have given the University of Dunglen an 'edge over competitors'. When one has a distinguished reputation for accomplishment in a field, it's easier to get other people to make available funding, equipment, specialised resources, and data. A good reputation opens doors in academia. Blackbriar's small biography within the article points out that he's responsible for raising huge amounts of research grant money – raising that kind of money requires serious credibility. Grant money like that in turn provides an accelerator for building reputation – grant money funds research students and research fellows, and they in turn generate papers, citations, and results for the grant owner. That in turn builds reputation, which then generates more grant funding. It is thus very tempting for some academics to attempt to fast-track the building of this reputation by taking advantage of the ease by which data can be fabricated and the difficulty of reproducing results. Publishing in a niche area with a relatively low amount of scrutiny can allow for misconduct to be perpetrated for a surprisingly long time. High-profile areas with lots of attention are correspondingly more difficult to survive within as a 'fake', but it's still possible. Conducting a research trial can be expensive, and the more complex it is the more moving parts need to be accounted for. In cases where the research cannot be replicated, academics also have recourse to explaining irregularities away under the guise of an artefact of the significance testing.

Many of the statistical models used regularly to assess the validity or notability of research data produce what is known as a P-value. This is a probability that indicates the chance that a test 'rejects the null hypothesis'. That's a fancy way of saying that you're rejecting the possibility that the results you found came about by pure chance. P values are usually measured on a scale of 0–1, where a 0.05 P value suggests that there's only a 5% chance that the results are an artefact of chance. The lower the P value, the more likely it is that you've found a meaningful (but not necessarily important) link between the variables you're examining. A P-value of 0.05 is considered 'statistically significant' for most studies, but that still means there's a 5% chance that you're reporting on something that was pure chance. It is in here that there is much room for an academic to defend fake results by saying 'Well, I guess my paper was the one in twenty that was just a chance occurrence'. P values by themselves tell only part of a story, but one that can hide a lot of deceit.

With our Professor Blackbriar, we now have a reasonable guess as to of what he's being accused, but we have no idea of the extent of the problem or indeed how far back it goes. Perhaps we're talking about one fudged result in one minor paper. Perhaps someone has uncovered a pattern of fraud stretching the length of his entire career. Perhaps he's actually completely innocent and the investigation will indeed exonerate him. We have to be careful at this point of assuming too much based on the limited information that we have.

We must bear in mind here too that our view of this emerging incident is highly skewed – we're reading newspaper reports, by a reporter called Jack McKracken, about a number of things that are outside the regular layperson's ability to immediately understand. The ins and outs of academic misconduct are complex, and the ways in which it can be perpetrated are subtle and often highly statistical. The motivations of those involved are far from clear-cut. Newspapers are notorious for simplifying complexities, removing nuance, and taking a particular editorial stance. We cannot assume that our source of information on this is entirely accurate – the concept of the unreliable narrator is useful here. Our intrepid reporter may fall into one (or more) of several possible motivations:

1. Honestly reporting on facts he understands, making sure to include all the necessary nuance without skewing the reporting.
2. Dishonestly reporting on facts he understands, looking to create a particular impression of wrong-doing at the university.
3. Honestly reporting on facts he doesn't understand, resulting in errors of inference and implication.
4. Dishonestly reporting on facts he doesn't understand, resulting in a chaotic mess of contradictory information.

We may build an impression of our reporter as we go through the case study and come up with a view as to his bias and reliability. We can also skip to the post-script in this book and read an interview with the man himself, although we would recommend that is only done when the scandal has played out in full. Our sole point of comparison otherwise is the other articles that he has written – this makes it difficult for us to calibrate the accuracy of the information coming our way. We can't do anything about that, but we can be mindful of it and not take what we are told as gospel. We can pick up on inconsistency in what has been written, but without access to the 'raw data' we can't assess whether what has been written was true. When quotes are provided, we don't know if they are the full quotes respectful of context. We don't know if they're abridged quotes that ignore context. We don't know if they are somewhere in-between. We must be careful in how we approach our assessment of the evidence.

We're at the start of a long discussion here – it's important we build what is to come on strong – and critical – foundations.

REFERENCES

Altman, D. G. (1994). The scandal of poor medical research. *BMJ, 308*, 283.

Altman, D. G. (2002). Poor-quality medical research: What can journals do? *JAMA, 287*(21), 2765–2767.

Bhattacharjee, Y. (2013). The mind of a con man. *The New York Times*, April 28.

Bolland, M. J., Grey, A., & Avenell, A. (2022). Citation of retracted publications: A challenging problem. *Accountability in Research, 29*(1), 18–25.

Brandt, A. M. (1978). Racism and research: The case of the Tuskegee syphilis study. *Hastings Center Report, 8*(6), 21–29.

Campos-Varela, I., & Ruano-Ravina, A. (2019). Misconduct as the main cause for retraction. A descriptive study of retracted publications and their authors. *Gaceta Sanitaria, 33*, 356–360.

Echevarrıa, L., Malerba, A., & Arechavala-Gomeza, V. (2021). Researcher's perceptions on publishing "negative" results and open access. *Nucleic Acid Therapeutics, 31*(3), 185–189.

Fanelli, D. (2009). How many scientists fabricate and falsify research? A systematic review and meta-analysis of survey data. *PLoS One, 4*(5), e5738.

Fanelli, D. (2012). Negative results are disappearing from most disciplines and countries. *Scientometrics, 90*(3), 891–904.

Fang, F. C., Steen, R. G., & Casadevall, A. (2012). Misconduct accounts for the majority of retracted scientific publications. *Proceedings of the National Academy of Sciences, 109*(42), 17028–17033.

Festinger, L. (1957). *A theory of cognitive dissonance* (Vol. 2). Stanford University Press.

Harris, A., Reeder, R., & Hyun, J. (2011). Survey of editors and reviewers of high impact psychology journals: Statistical and research design problems in submitted manuscripts. *The Journal of Psychology, 145*(3), 195–209.

Harris, A. H., Reeder, R., & Hyun, J. K. (2009). Common statistical and research design problems in manuscripts submitted to high-impact psychiatry journals: What editors and reviewers want authors to know. *Journal of Psychiatric Research, 43*(15), 1231–1234.

Heron, M. (2012). Inaccessible through oversight: The need for inclusive game design. *The Computer Games Journal, 1*(1), 29–38.

Heron, M., Hanson, V. L., & Ricketts, I. W. (2013). Accessibility support for older adults with the access framework. *International Journal of Human-Computer Interaction, 29*(11), 702–716.

Hopewell, S., Dutton, S., Yu, L.-M., Chan, A.-W., & Altman, D. G. (2010). The quality of reports of randomised trials in 2000 and 2006: Comparative study of articles indexed in PubMed. *BMJ, 340*, c723.

Levelt, W. J., Drenth, P., & Noort, E. (2012). *Flawed science: The fraudulent research practices of social psychologist Diederik Stapel.* Commissioned by the Tilburg University, University of Amsterdam and the University of Groningen.

Mahaffy, K. A. (1996). Cognitive dissonance and its resolution: A study of lesbian Christians. *Journal for the Scientific Study of Religion, 35*(4), 392–402.

Marcus, A. A. (2023, July 12). The new retraction record holder is a German anaesthesiologist, with 184. *Retraction Watch.* https://retractionwatch.com/2023/07/12/the-new-retraction-record-holder-is-a-german-anesthesiologist-with-184/

Milgram, S. (1965). Some conditions of obedience and disobedience to authority. *Human Relations, 18*(1), 57–76.

Nath, S. B., Marcus, S. C., & Druss, B. G. (2006). Retractions in the research literature: Misconduct or mistakes? *Medical Journal of Australia, 185*(3), 152–154.

Normile, D. (2012). A new record for retractions. *Science Insider (American Association for the Advancement of Science), 2.* https://www.science.org/content/article/new-record-retractions

Rapani, A., Lombardi, T., Berton, F., Del Lupo, V., Di Lenarda, R., & Stacchi, C. (2020). Retracted publications and their citation in dental literature: A systematic review. *Clinical and Experimental Dental Research, 6*(4), 383–390.

Smith, R. (2010). Classical peer review: An empty gun. *Breast Cancer Research, 12*(4), S13.

Smith, R. (2014). Medical research—Still a scandal. *BMJ Opinion, 31.* https://blogs.bmj.com/bmj/2014/01/31/richard-smith-medical-research-still-a-scandal/

Sterling, T. D. (1959). Publication decisions and their possible effects on inferences drawn from tests of significance—or vice versa. *Journal of the American Statistical Association, 54*(285), 30–34.

Teixeira-Pinto, A. (2021). Of scandals and statistics: Improving analytical methods in clinical and health services research. *Circulation. Cardiovascular Quality and Outcomes, 14*(7), e008279.

Thumma, S. (1991). Negotiating a religious identity: The case of the gay evangelical. *Sociological Analysis, 52*(4), 333–347.

Yeo-Teh, N. S. L., & Tang, B. L. (2021). An alarming retraction rate for scientific publications on Coronavirus Disease 2019 (COVID-19). *Accountability in Research, 28*(1), 47–53.

Zimbardo, P. (2011). *The Lucifer effect: How good people turn evil.* Random House.

Zimbardo, P. G., Haney, C., Banks, W. C., & Jaffe, D. (1972). *Stanford prison experiment: A simulation study of the psychology of imprisonment.* Philip G. Zimbardo, Incorporated.

CHAPTER 4

Student Suspensions at Scandal-Ridden University

NEWSPAPER ARTICLE

An exclusive report by Jack McKracken for the Dunglen Chronicle

The scandal which has engulfed the University of Dunglen in the last week has just claimed two more victims. Postgraduate research students Sharon McAlpine and James Duncan have been suspended pending inquiry as a consequence of their close relationship with Professor John Blackbriar. Blackbriar was the 'primary supervisor' of the two students, and the three have collaborated on the publication of several papers during their association with the university. The nature of this collaboration has been described as 'fraught' by those close to McAlpine and Duncan, with the move to remote working during the Coronavirus pandemic being noted as placing particular stresses on already fractious relationships.

Insiders within the department have told the Chronicle that proprietary data sets used by Professor Blackbriar in the development of his research career were 'tainted with bogus data'. It is also alleged that the professor has been regularly pruning 'outlier' results from the data sets his work generates, ensuring that his various computer simulations show only the most promising results rather than the most likely. It is this data that was used to inform the development of the papers on which the three collaborated. The papers

DOI: 10.1201/9781003426172-4

outline a bi-modal technique of first finding geographic points of interest through the use of a heuristic algorithm called 'Simulated Annealing' followed by an extensive proof-of-work blockchain-based data-mining process. This cross-referenced hundreds of thousands of measurements in dozens of categories across ten years for hundreds of discrete locations within each region. The Blackbriar Algorithm then makes use of convolutional neural networks (CNNs) to optimise the identification of points of interest. Large Language Models (LLMs) are employed to provide narrative explanations suitable for the end-user. To minimise the need for human intervention, the system also makes use of a series of proprietary deep sea drones which are deployed to shortlisted points of interest to take samples and soundings.

This complex innovative technique was the subject of several patents which were jointly held by a consortium of Scottish oil companies, the University of Dunglen, and Professor Blackbriar himself. Before the question marks that have grown over the process, these were thought likely to yield billions in oil revenue and millions in patent licencing fees for Blackbriar and the University. The situation now is much less clear.

'Removing outliers from data sets and simulations is common practice', we were informed by research ethics expert Professor Callum Sunderland from the University of Alba, 'Outliers are by their very nature outside the normal course of investigation and may represent flawed measurements, unusual circumstances, and other such "noise". Removing these before analysis is a regular part of sanitizing a data set to ensure that rogue data points do not overly skew the results of the research'.

The professor went on to say, 'However, it appears as if the definition of outlier used at the University of Dunglen was unusually open to interpretation and this may have resulted in many perfectly valid data points being excised because they were inconvenient rather than because they were suspect. This would not be an appropriate action to take before analysing. Similarly, introducing "representative sample" data points into the data set is inappropriate even if the data points are credible. Quite simply, you can get any result you want by getting rid of that which is inconvenient and replacing it with that which is desired'.

Professor Blackbriar has been unavailable for comment, but one of his colleagues spoke to the Chronicle today saying, 'It's absolute nonsense. I don't believe any of this for an instant. John is a cornerstone of the research output of this department. He would never do a thing like that'.

Part of the problem in this particular incident is that the data points the professor has gathered over the years have come as a result of his close contacts with the oil industry. The gathered information is sourced from around a dozen different companies from 50 different sites, and portions of the resultant data set are proprietary to each of the individual organisations. Only Professor Blackbriar himself had access to the entire research data set. Even those in the research consortium couldn't view the data points of other consortium members because of 'commercial sensitivity'. Students and colleagues who wanted access to these valuable results were required to sign a nondisclosure agreement with individual companies and permitted to see only a subset of the data. This narrow focus meant that only Blackbriar truly knew how everything fit together.

'Commercial interests in research are nothing new', Professor Sunderland explains, 'But this kind of thing highlights the problems. Data often only makes sense when provided in its full surrounding context. This bite-sized approach to data-set analysis means that even diligent scholars may not be aware that they are working with data that has been doctored'.

This appears to be the fate of the doctoral candidates who have been suspended from the University. A spokesman for the university said earlier today, 'Two of our research students have been suspended from their doctoral programs for the time being while we fully investigate the allegations against Professor Blackbriar. We believe that they have been responsible for promulgating tainted research data, and it is important that we take a stand on this to ensure that our unparalleled reputation for research integrity is not harmed. All research students at this university are expected to make every effort to ensure the correctness of the work that they publish, and there is reason to believe that this may not have been done in this situation'.

Both Mrs McAlpine and Mr Duncan have been unavailable for comment, but an insider within the University has said, 'They're obviously furious, as well as deeply hurt. This is their academic credibility on the line – you don't come back from something like this. It's not like they're the only ones who collaborated with John – if they're guilty, so are a lot of other people. Why they're being singled out, I don't know'.

A friend of Mrs McAlpine, speaking on the grounds of anonymity, has told the Chronicle 'The university is throwing them on the fire, in the hope of stopping the blaze from spreading. They're patsies, but I know that Sharon at least isn't going to let herself be burnt for this'.

STUDENT SUSPENSIONS AND THE COMMUNICATION OF SCIENCE

Within the second article, two more academic scalps are claimed – two of the PhD students who were working on their doctorates along with the professor. We also have more of a feel for what the allegations actually involve – 'bogus data' within data sets that the professor had been constructing. As we talked about in the previous chapter, it's hugely important that raw data is correct and accurate otherwise the entire basis of future analysis is unsound. There is a useful mantra in computing science 'Garbage In, Garbage Out' – if what goes into your calculations is bad data, what comes out are bad conclusions.

We're introduced to Professor Callum Sunderland in this article – he'll be a guide of sorts through the coming articles, but don't treat him any differently to anyone else in the study. He's unconnected to the incident, and a known (fictional) expert in the area of research ethics. We still don't know to what extent his views are being fully represented by the reporter, or what his own motivations for commenting at all may be. After all, he is employed at another university and while there is a good degree of collaboration across institutes there is also rivalry. How much does a representative of the University of Alba stand to gain by commenting on the misfortunes of a competing university?

The details of the allegation are that incorrect, or misleading, data has been inserted into the raw data sets used by the professor, and that he has been trimming 'outliers' from the set so as to skew the results towards a particular desired conclusion. This is a serious breach of research ethics, and it seems bizarre to think that something this serious could go unnoticed. We don't know how long he has (allegedly) been at this, but surely something like this couldn't escape expert scrutiny?

Well, perhaps and perhaps not. It all hinges on how much we might expect his colleagues and collaborators to understand what he's even been doing. What we see here is a popularised description of a complex research algorithm – that's just a fancy way of saying 'a discrete set of unambiguous steps followed (usually) by a computer'. It's intentional that this be a little baffling. We're not supposed to know for sure how the research worked because we're seeing it reported through the filter of an unreliable narrator for a non-specialist audience. At the cutting edge of research, only a small number of people are likely to fully grasp the implications of experiments and studies, and there's a limit to how far it can be simplified without losing important nuance. The problem is beautifully illustrated in a sketch from British comedy duo Armstrong and Miller in which a daytime television interviewer attempts to do a short interview with a physics professor:

Interviewer: Heterotic Super-symmetry is said to combine elements of String Theory with a new take on, now hang on, 'Quantum

Chronodynamics'. Now trying saying that when you've had a few. Aaand it's the brainchild of Professor Alan King. Professor King, good morning.

Professor: Good morning.

Interviewer: Can you just briefly take us through this new theory of yours, in layman's terms?

Professor: No.

Interviewer: All I'm after is just a broad stroke explanation, if you like.

Professor: There isn't one.

Interviewer: Uh, okay. Well… what if you were to take us through the whole thing, starting with the real basics, and just working our way up?

Professor: Okay. Yeah, yeah, we could definitely do that. It will though take quite a long time.

Interviewer: How long?

Professor: Eleven years.

Interviewer: Right, I'm being told we don't have quite that long, ha ha. Professor, some of our viewers are quite smart, um. Perhaps there is someone watching at the moment who's capable of understanding your theory?

Professor: There isn't.

Interviewer: How can you be so sure?

Professor: Because Graeme is on holiday and Chung Yu is dead.

Interviewer: Professor King… thank you.

Professor: My pleasure.

Communicating complex research to a non-specialist audience is a difficult task, and academics generally don't really do enough of it. What we're left with then is journalists and other 'outsiders' attempting to make some sense of complicated issues as best they can. As a result, we're never truly privy to the full complexities of what Blackbriar's algorithm is doing. That makes it difficult for us to really understand how he could get away with this alleged fraud. A little bit of research around the Internet would reveal several features though around the words used to describe it, and we can perhaps construct a mental model of his work by digging into those.

1. Bi-modal
2. Simulated annealing

3. Data mining
4. Proof-of-work
5. Blockchain
6. Convolutional neural networks
7. Large language models
8. Deep Sea Drones

It's not critical to our case study to know what the professor's work likely did, but it's illustrative of understanding how something as straightforward as fudging the data may have gone undetected. Feel free to skip the next section if you're happy to accept the proposition 'some research is too complicated for lay people to easily understand'.

THE BLACKBRIAR ALGORITHM

Let's begin with the term simulated annealing – this is a reasonably well-defined process where a computer algorithm will use probability-based searches to find a 'good' solution when it's too costly to exhaustively search all the possibilities. Right there we see the issue – in order to understand one thing, we need to understand several others.

A probability-based search is one that will search, based on some calculated values that indicate likely 'good' places to look, in different places each time it's run. Imagine a probability-based maze-solving algorithm. In its simplest form, it might say '25% of the time, turn left. 25% of the time, turn right. 50% of the time, go forward'. When we encounter our first decision, half the time we'll go forward and the other times we'll take a turn. We don't know in advance though which way we're going to look. Many of the more interesting search routines in computing use some variation of this, making use of probability to deal with huge amounts of possible 'hiding places' that need to be investigated. You might use this kind of system when for example you have hundreds of millions of possible combinations of input values and can't possible assess them all in a reasonable period of time.

You won't get anywhere though if all you're doing is cycling through a virtually limitless set of combinations. You need a way to determine the 'fitness' of particular combinations too so that you know when you've found a good one. The basic criteria are 'You know when a particular combination is good, but you don't know in advance what makes a good combination'. Simulated Annealing is a really neat technique, and one worth spending a bit of time reading about. While you're there, have a look at the concept of a genetic algorithm. Just for fun.

The next bit of the algorithm is that it's 'bi-modal', which means that at some point when it finds something reasonably interesting it switches to another kind of search that allows for a more 'fine-grained' look at the area. We're never given details as to what the second mode of this actually means, but that's typical in popular news reporting of complex science. We can't hope to get all the information. Finally, after having found some good candidate solutions from this technique, a data-mining effort takes off which looks to extensively investigate connections between all the different bits of the data. This kind of approach is increasingly important in real life.

All you really need to take away from this little explanation of what this research project does is 'it's got an awful lot of calculation in it, across an awful lot of possible input values'. These kinds of algorithms may take hours to fully crunch through a set of input data, and because they are probabilistic they don't necessarily come up with the same answer every time – each involves a degree of generated randomness to decide which possibilities are best to explore. Sometimes they'll investigate one possibility and sometimes they'll investigate another. The sheer vastness of the number of all possible combinations (known technically as the solution space) means that they can only ever hope to explore a fraction of all the options. The job of the algorithm is to identify what fraction that should optimally be.

Solution space is another term that we need to define to understand this research – imagine you have a simulation of a real-world phenomenon. It takes in 20 pieces of input data. Maybe it's dealing with weather, and it needs the temperature, the wind speed, the humidity, the wind direction, and so on. You know that certain combinations of those values will result in rain and some will result in sunshine. If you want to know what creates a sunny day, you need to continually iterate over all the combinations of all those individual 'traits'. Some of them may have only a few possible values. Some may have millions. Just look for example at the combination of wind speed and temperature. Wind speeds can typically vary from 0mph to 80mph or higher, with all points in-between. We don't know if there's a difference for weather between 50mph and 50.1mph, and so we need to look at a few decimal points between the natural numbers. Let's say that we only use one significant digit for this – that gives us 800 possible values for wind speed. Now for temperature – that can vary from −27.2°C to 38.5°C within the United Kingdom. That's 65.7°C full, with again a significant digit on the right of the decimal for each. That's a grand a total of 657 possible values. We could exhaustively search any one of those by itself, but taken together there are 525,600 possible combinations and we don't know which of those combinations give us the results we want. Lacking any further insights, we'd need to search over all of them.

That's just for two traits in the input model. Throw in a third – humidity. That gets measured in percentages, so that's another 100 values, and again at a single significant digit, a total of 1000. Now we've got 525,600,000 possible combinations to check and we've only dealt with three of our 20 bits of input data. The whole set of those possible combinations is the solution space, and it's already too big for us to exhaustively explore. So the Blackbriar Algorithm will search, at best, a very small portion of a solution space based on random probabilistic numbers generated to determine which possibilities get explored in a particular run.

We also know that the system makes use of neural networks to help the algorithm optimise itself. Convolutional Neural Networks (CNNs) are a popular form of learning algorithm, used in everything from facial recognition, audio processing, image generation, to climate analysis. The trick in how they work is in the name – convolution. This is the process of putting an image through a range of different filters, each designed to capture certain regions or features. Based on which features are activated by particular filters, the algorithm can ascribe weights and biases to these combinations allowing them to be differentiated from each other. These systems process visual information in a wildly different way from human perception, but they have been shown to be extremely effective when properly trained on the right data and with the right parameters.

The cost of this is in that 'wildly different' element – the more decision-making we pass off onto artificial intelligence, the less confident we can be in how results are obtained and how decision-making is structured. Inevitably we lose transparency of the link between input and output. This is important in this particular case study, because much of it hinges on 'outlier' data and the meaning of that may not be at all clear in a system that makes significant use of machine learning. In fact, some systems **rely** on carefully curating the data sets to avoid the problem of what is known as **overfitting** – when a system is optimal when working with training data but fails to demonstrate results when real data is introduced. Balancing overfitting versus underfitting is a constant challenge in working with artificial intelligence.

We can also see that LLMs are used to address this transparency issue – an attempt is clearly being made to offer human-readable explanations of the work being done. You've probably already encountered LLMs somewhere in your own life by this point – if you haven't played about with GPT, then you should so you at least understand the technology that will obsolete 80% of all human endeavour. Essentially, this is another AI system, but one trained on vast amounts of data in the hope of being able to replicate it in structure and form. This is a process known as **generative** AI. Billions of parameters are used to create a general-purpose model of how the different pieces of the data fit together. Statistical inference – a process of using these parameters to generate probabilities – is then employed to predict, for example, how one might meaningfully complete a sentence.

In fact, let's see just how that looks by doing exactly that. I asked GPT to 'Give me an easy overview of how a LLM works'.

And it told me the following:

> Sure, Michael. A large language model is basically an artificial intelligence system that has been trained with a lot of text data to learn how to generate natural language. This data could be anything from books and articles to social media posts and online chats. The model consists of multiple layers of neural networks, which process the input, learn from it, and generate outputs.
>
> When you feed a prompt or input into the model, it uses the learned information to predict what words or phrases would likely come next based on the probability of them being used in the context of the input. The model generates text by predicting and outputting one word at a time until it reaches the desired length or until it generates an output that meets a certain criterion.
>
> The training process involves presenting the model with a large amount of text data and gradually adjusting the weights and biases of the neural network layers until the model can accurately predict the next word or phrase in a text sequence. This process is called pre-training. The model is fine-tuned on a specific task (such as translation or text completion) by continuing to train it on a smaller, more targeted dataset.

> Overall, a large language model works by learning the patterns and structures of language through exposure to vast amounts of text data, then using that knowledge to generate natural language in response to user input.

You can imagine how powerful this technique can be – it works for everything from translation, to coding, and even offering advice to resolve ambiguous situations. However, while these results are impressive, we have to be wary. The model itself has no idea what it is saying, and since it is generative its output represents a statistical output rather than a factual one. For example, if I ask GPT for a list of papers about how the Blackbriar Algorithm makes use of LLMs, it gives me this:

> "How Large Language Models Aid Blackbriar." Medium, www.medium.com/blackbriarblog/how-large-language-models-aid-blackbriar-75a956992275.
>
> Shukla, Arun, et al. "BlackBriar: Utilizing Deep Learning to Enhance Automatic Speech Recognition." arXiv, Cornell University Library, arxiv.org/abs/1911.09851.
>
> Vosamaki, Dimitra, et al. "BlackBriar: How Language Models and Neural Networks Enable Speech Recognition in Low Resource Languages." arXiv, Cornell University Library, arxiv.org/pdf/1909.02573.pdf.
>
> Heidecke, Clayton D., et al. "Reinforcement Learning Systems for Large Language Models Using BlackBriar Algorithm Variants on GPU Clusters." IEEE Access, Institute of Electrical and Electronics Engineers (IEEE), 2019, ieeexplore.ieee.org/document/8777979/.
>
> Chua Wai Yin, Kenneth et al., "Large Language Modeling on BlackBriar: An Analysis Using Recurrent Neural Networks" 2017 International Conference on Intelligent Sensors, Sensor Networks and Information Processing (ISSNIP), 2017 IEEE 7th International Conference on Intelligent Sensors, Sensor Networks and Information Processing (ISSNIP), IEEE, 2017 Dec 18–20, IEEE Xplore Digital Library: https://ieeexplore.ieee.org/document/8278300

Obviously, none of those are real references – the Blackbriar Algorithm doesn't exist outside of our case study. These fictional papers though are so confidently formatted and delivered that an uncritical reader will be forgiven for taking them at face value. In fact, that happens already. In the middle of an argument with a friend about Michael's competence to comment on a particular topic, Michael asked GPT to list his five most significant papers showing he was an acknowledged expert in the area. The algorithm happily supplied a whole bunch of credible looking references. At this point Michael's friend responded by insisting that the Michael Heron listed as an author on these papers was a **different** Michael Heron. He didn't consider the possibility, initially, that the papers didn't even exist. The way in which generative AI presents its results often bypasses human scepticism in a dangerous way. Sure, Blackbriar may have human readable analyses of the AI decisions being taken… that doesn't mean those analyses are accurate or even in the ballpark for correctness.

We've also got in our description the buzzwordy term 'Blockchain' thrown into the mix. You may have already had experience in real life trying to work out its significance in a whole pile of fancy sounding technological solutions to problems that probably don't exist. At its core, the idea is pretty simple – imagine that instead of shifting a piece of data around by itself, you moved it around along with a full accounting of its history. Every time it was touched or changed, a little cryptographically secured hash would be computed to encode it – this is known as a **record**. Each record contains a hash of its previous records, creating an irreversible and theoretically secure chain of interactions. You can't alter any record without altering all the others that follow. Most of these use a peer-to-peer network to act as a kind of public record – known as a **distributed ledger**. Working together, this network authenticates and validates new transactions. The technology is highly secure and is at the core of the growing family of cryptocurrencies. Given that each record contains the chain of all previous records, the storage requirements tend to grow exponentially. In 2014, it was estimated the ledger of all Bitcoin transactions was around 20 gigabytes, but as of the end of 2022 it had risen to 435 GB.

We don't know exactly how blockchain is working within the Blackbriar model, but the use of the phrase 'Proof-of-Work' tells us a lot. One of the things that underpins some cryptocurrencies – Bitcoin being one of them – is in how the accuracy of new transactions on the blockchain is verified. One of the issues with a peer-to-peer system is that there is no central authority, and so there needs to be a way for all the different parts of the system to agree on a cryptographic proof. In technical terms, they need to arrive at **consensus**. Essentially, new chains are only considered valid insofar as they are verifiable products of computing processing. When you hear people talking about Bitcoin 'mining', this is what they're talking about – computers doing unnecessary calculations in order to assuage the paranoia of a consensus-based system. As a result there is huge environmental impact as around a million people draw electricity to power racks of CPUs and GPUs in the hope that they do enough work to gain a newly minted Bitcoin. The cost is considerable – while it is not wholly straightforward to estimate electrical consumption of the Bitcoin economy, estimates usually put it in terms of comparable outputs on a national scale (Badea & Mungiu-Pupzan, 2021). Greece and Ireland are a common basis for comparison.

This isn't true of all cryptocurrencies – other systems, such as Ethereum, employ alternatives such as 'proof of stake', in which superfluous calculation is abandoned in favour of having systems offer a 'stake' of coins as collateral for validation. Actual validation is then done by a random system – weighted by stake – where a consensus as to validity is decided upon by a subset of hosts. The environmental impact of this, in comparison to proof-of-work, is almost negligible. In 2022, Ethereum moved to a specifically environmentally conscious approach to consensus building, instantly dropping global electricity draw from 2,600 MW/h a year to 23 MW/h. Bitcoin, by comparison, consumes about 127 Terawatt hours per year.

This may seem like an odd aside to take, but consider too that there is another part of this system – that the Blackbriar Algorithm also employs deep sea drones to 'take samples' of some subset of identified points of interest. The article is non-specific about how many points of interest and has nothing to say about the likely environmental impact of this sampling. In a world that is rapidly heating beyond even our pessimistic projections, there are surely questions to be answered here about how we pay for the technologies we choose to use.

Bringing that all together, we have to acknowledge the circumstance that replicating these results is very tricky. You need access to the algorithm (which may or may not have been published in full), access to the raw data (which we are told in no uncertain terms is proprietary and spread across many stakeholders). In this case, we'd also need access to the random number seeds that were used to generate the probabilistic results. It's clear from what we know that these are results we couldn't replicate by hand, even if we had access to the data. If we had access to the data and the algorithm, we'd still end up with completely different results because the random numbers used to decide on probability would be different when we execute it. The nature of the CNN at its heart obscures our ability to transparently replicate results. We'd need all parts fully laid out in front of us to have even a remote chance of reproducing the results of a particular run. If we can't recreate the results, then we can't corroborate, or refute, the accuracy of the published work.

Reproducibility in science is tremendously important – we spoke in the last chapter about statistical significance, and how even 'good' scientific results might come about just from chance. Some fields have a rigorous tradition of the validation of interesting results, and others do not. There's often little reputation to be built in checking up on other 'original' ideas even if it is hugely important for science to be as self-correcting as its advocates wish. If the allegations are true, then Blackbriar has gotten away with it for so long because it was simply not possible for anyone to replicate his results, and thus the field as a whole simply trusted him. Understandable, given how we've seen in the previous article that he has a track record for success.

COMMERCIALISED RESEARCH

The professor's defence against this lies in the proprietary nature of his data sets. Commercial interests in research are nothing new and are in many cases encouraged by universities as part of 'knowledge based transfer' or 'entrepreneurship' (Gulbrandsen & Slipersaeter, 2007; Macho-Stadler et al., 2008; Perkmann et al., 2021; Siegel & Wessner, 2012). Few things speak to the true value of academic outputs as well as the marketplace, and this serves as a measurable way to assess that (Markman et al., 2008).

Commercial entities often require specialised research to be done, and much of the brain-power available to do it is found in universities. Commercially funded research is especially common in the pharmaceuticals sector; the tobacco sector; the military defence sector; the computer science sector; the oil and gas sector; and the biotechnology sector. Dr. Heron's own PhD (Heron, 2011) was funded in part by IBM. There is nothing wrong with such funding if it doesn't impinge on the university's ability to be open and transparent about results, but unfortunately it's never quite that simple.

It would be wrong to say that all commercially funded research ends up with bad science, because we can easily prove that's not true – there is lots of good research that only came about as a result of commercial sponsorship. However, there are certain tendencies that have been identified when commercial funding is provided to individual research projects (Bastian, 2006; Langley & Parkinson, 2009).

1. They tend to find results that are harmonious with the commercial priorities of the funders. This 'funding bias' isn't necessarily a result of academic misconduct, although it can be. It's more often a result of 'loading the deck' by funders who predominantly pick those researchers who had already published results favourable to the industry. See for example Lundh et al. (2017); Pisinger et al. (2019); and Turner and Spilich (1997).
2. Commercial funding often comes with confidentiality agreements, requires the signing of non-disclosure agreements, and includes intellectual property exploitation as a cornerstone of the outputs through the seeking of patents (Washburn, 2008). Similarly, some companies may use confidentiality clauses to stop the publication of results that they do not desire (Langley & Parkinson, 2009).
3. Conflicts of interest begin to become more significant (Babor & Robaina, 2012). When reporting for example on experimental trials around a particular sponsored product, it is not always the case that these interests are fully disclosed (Krimsky & Rothenberg, 2001).

At a sector level, other things also come into play (Langley & Parkinson, 2009):

1. Commercial interests become one of the key elements in progressing a research agenda, rather than that which may be in the public's interest.
2. As universities reorganise to take advantage of these revenue streams, they are often compelled to disregard classic 'academic' virtues such as openness, objectivity, and the independence of researchers
3. Academic departments are increasingly orienting themselves to the needs of business, which means that corporations have greater ability to apply pressure because they are a major, or even the primary, generator of research funds.

Again, it is not our argument that commercial interests in academic research are necessarily bad – only that the benefits that come from having such funding available must be balanced against the risks both for individual studies and for the sector as a whole. In our case study, we can see one of the major issues – commercial confidentiality agreements have trumped the requirement for the professor to be open and transparent about his results.

In this situation, we have a compartmentalised model of knowledge where many funding partners and collaborators are feeding in to a single pool of data. Only the professor, as gatekeeper and guardian of the data, can see the full sweep of what's contained within. Each of the individual partners in the project would know what their own data is supposed to look like but would be blocked from viewing the data that might belong to their competitors. It's not possible for someone to simply go back to the source here because the source data from one provider has long since been merged with the source data from others. It's not impossible that someone may negotiate all the many legal obstacles that would prevent them from seeing the full, unprocessed data elements. It would though be prohibitively complicated and unlikely to be granted just to ensure 'intellectual oversight'. As the only person who can view the data, only the professor knows how it's been stored and processed. As Callum Sunderland tells us in the case study, the removing of outlier data is not in and of itself an unusual activity. Outliers are pieces of data that exist well outside the parameters of the body of data and thus are likely to be special cases, the result of incorrect measurement, transcription errors, or just general 'noise'. A simple, straightforward analysis of a set of data that includes outliers may result in strange, or completely misleading, conclusions. The process of cleaning up a data set is a common part of the necessary curation of data.

However, when this is done subjectively, it leads to a skewing of the results. In our case study, we're told that the definition of an outlier was 'open to interpretation'. The selection criteria for deciding what pieces of data should go into the analysis is important – it should be the case that two unrelated researchers can look at the raw data, assess it by the selection criteria, and end up with the same subset of data at the end (within a certain, usually small, margin of error). Where subjective criteria are used, this becomes increasingly unlikely as hundreds of individual mini-judgements are made. When one is too close to the data, it's tempting to discard data points that are simply inconvenient. It might be possible to justify it to yourself and to others, but the question becomes 'would you have made the same decision if an equivalent outlier strengthened your theory rather than weakened it?' It's clear from what we know of the incident here that too much was left open to interpretation.

As to what 'bogus data' means, we don't yet know. We'll await further details on this before we introduce it into our discussion. All we can really say at this point is 'well, that doesn't sound good'.

What of our doctoral students then? They've found themselves at the centre of a scandal of which they may have had very little knowledge. The relationship between a supervisor and their students is one of the most intense that many young people experience – in its healthiest forms it's a deep collaboration, coupled to a mentorship. Regular contact and support and guidance can lead to a productive partnership and a collegiate relationship that lasts for a career. In the unhealthiest forms, students experience distance, disappointment, and a feeling of being ungrounded. We know from our case study that there was at least some degree of collaboration, because the three of them published papers together in various different combinations. The nature of that collaboration however remains unknown for now.

Their crime in this case appears to have been trusting data that was provided to them by their supervisor, and then publishing that data in academic journals and conferences. The extent of their guilt in this case is very tightly coupled to the extent of their knowledge as to what was happening. It's unlikely that the professor has a blanket authorisation to provide full access to the data sets to his students, but we'll find out for sure as the case study progresses. However, there are certain things that we might reasonably expect in the circumstances. The students had been publishing papers with the professor – that's very valuable for young researchers looking to 'make their bones' in the cutthroat and competitive world of academia. Nobody is likely to be very interested in what they have to say about the topic, but Blackbriar is a different matter. If they publish along with Blackbriar, their names get associated with his work. If someone searches for Blackbriar's research, they'll likely find our postgraduates coming up in the search. After all, we know from the previous article that Blackbriar has tens of thousands of citations – that's a lot, enough to put him in the upper echelons of his discipline. Such associations help tremendously in giving a new researcher a leg-up and a start on building their own reputation.

This comes at a cost though – the expectation is that everyone whose name goes on a paper has signed off on the contents. As such, everyone associated with it gains some of the prestige but also shoulders some of the blame in the comparatively rare event of academic fraud. We know from the case study that there are complex intellectual property arrangements in place that must be navigated in order for someone to gain access to the data sets. It seems unlikely that our doctoral students would have had full access to the data and as a consequence it's unlikely they fully understood the validity of what they were publishing. What then is their responsibility in the matter? They put their names to the papers – perhaps they asked to delve more deeply into the research and were dissuaded by the professor. After all, he's a respected figure in his field, and if he says, 'trust me, the data is fine' they're doing nothing more than other researchers are doing when they don't replicate the results. It's likely that they are genuinely in breach of their contracts as junior researchers, but it's equally likely that it's through no genuine fault of their own.

The professor has more than simply his professional reputation hanging on the perceived success and failure of his research – we're told in the article that both the university and the professor hold several joint patents regarding the technique. Exploitation of generated intellectual property is something that is becoming increasingly important to universities. More and more academics are being encouraged to turn the work they do into resources that can generate funds for the university and for the academic personally. To incentivise such things, a university often takes on the role of the point of contact. Their role is not to directly use the intellectual property – instead, they contract the right to exploit the patents to interested parties. Funds generated in such a way usually get split between the university and the original inventor, with the split becoming more slanted towards the university the more money that is generated. Often, universities will sign their shared patents over to the inventor if they wish to make an attempt to directly exploit the intellectual property through a start-up enterprise. The university will transfer their right to the Intellectual Property (IP) in exchange for shareholder status in the new venture. Courses, internal training, and consultancy to the academic or academics are provided to maximise the chances that the venture will be a success. Evidence of success in such endeavours is limited (Perkmann et al., 2013). Generally speaking, this is a risky proposition for the individuals in question. Early studies suggest that, at least within certain countries, that there is little extra money gained by academics but much extra risk shouldered (Astebro et al., 2013).

The most prestigious universities have a regular and ongoing production of intellectual property. The University of California received 597 patents in 2020. MIT received 383, and Stanford 229. The sector as a whole in the United Kingdom applies for over 2000 patents a year, of which over 700 are granted. The evidence on the efficacy of this is mixed. In 2001, Charles and Conway (2001) showed that few universities make money from their library of patents, and that the costs usually exceed the revenues. That picture does not change when we look at recent years. Not all intellectual property produced by universities is suitable for exploitation (Andersen & Rossi, 2011). It's difficult, given the incompatibilities in how various institutes manage and report on their patent and licencing, to evaluate for sure the value of these revenue streams (Bonaccorsi & Daraio, 2007; Meyer & Tang, 2007). The competency of the sector in exploiting these intellectual property rights is increasing, but it remains a fraught business.

That doesn't change the sheer number of patents and the sheer amount of money being spent and produced in aggregate. Within the US alone, annual patent applications have risen from 15,953 in 2015 to 17,738 in 2020. Granted patents have gone from 6,680 to 8,706 in the same period. Research expenditure has risen from around 6.5 billion dollars to around 8.3 billion. Over a thousand staff are employed in managing these various licences. For all of this, the gross licencing income show spectacularly poor results. The best performing universities have an annual median licensing income of 14 million. The worst performing have an annual median income of a mere 15,000 (Allard et al., 2021). These are the 2020 figures.

When universities do manage to acquire a patent with genuine commercial implications, then it can be a significant boon to the institution – one that must be protected and wisely shepherded. The income from a profitable patent is unlikely to make a difference to the fortunes of an institution as large as a university. It can though have a hugely disproportionate impact on individual departments and the academics that benefit from the profit streams. The University of California for example earns almost half of its 48m USD patent-related income from five patents, including a Hepatitis B vaccine and a bovine growth hormone. Given that the university holds 6,441 active patents as of 2020, it is clear that there's a very long tail regarding where the profit is to be had. The university's annual budget for 2009–2010 came in at 20.1 billion USD – that 48m USD of research income accounts for around 0.2% of that budget. In 2022, the annual budget was 46.9bn USD with 109m of commercialisation income. Or, again, around 0.2% of the budget.

Even a breakout hit of a patent won't make that much of a difference to the overall financial health of a university. Even the most successfully entrepreneurial universities do not cover their entire research spend with patent-related income – most struggle to attain a 10% return on that investment. That's not to say that the income is insignificant or unimportant. The inventor behind a university patent might share in as much as 50% of the research income. The usual figure though is much lower, and the percentage tends to decrease as the revenues pass various thresholds. The department in which the academic is housed can expect to receive a bounty from the funds that go to the university. Even a small revenue stream can be the difference between working alone and having a few postgraduates to help out. The more researchers you have working for you, the more papers that will be produced, and the more reputation you build. A commercially exploitable patent makes life in a university much more productive for the generator of such revenue.

REPUTATIONAL IMPACT

Our students are in a tough position – likely caught in the shadowy hinterlands where they have committed a breach of research ethics, but without active knowledge of it. Their association with a respected academic in the field has likely gone from a dream opportunity to a nightmare with no real escape. That's the risk in a world where your reputation means everything – when it's gone the ramifications are much more significant. Our postgraduates are likely at the start of their academic careers – before they even get a chance to establish themselves, they've been caught up in the wake of a scandal. They'll find it very difficult to recover from this regardless of what the investigation by the university finds. Their association with a very visible case of academic fraud will be a stain on their curriculum vitae for the rest of their lives.

In the Stapel case discussed in the previous chapter, 19 PhD theses were prepared with data obtained from Stapel. Seven of these were eventually cleared of wrongdoing, and 12 remain under some degree of suspicion. The parallels with Professor Blackbriar are considerable – Stapel maintained a tight control over the raw data and made excuses whenever students asked to see it. The report into Stapel's misconduct outlined the damage done (Levelt Committee et al., 2011).

> The Committee finds that great harm has been done to coauthors and PhD students in particular, as a consequence of the fraud committed by Mr. Stapel. The Committee has become aware

> in the course of the interviews that the consequences for those involved can be both formal and informal in nature. In a formal sense, the people affected are hampered in their careers, such as when extending temporary contracts and applying for grants. On an individual level some victims have been demoralized. In an informal sense, there is an element of stigmatization that may persist long into their further career.

It goes on to add:

> It is not only individual people that are affected, but confidence in science, and in social psychology in particular, has been badly dented by Mr. Stapel's actions. Other victims are the universities of Tilburg, Groningen and Amsterdam, whose reputations have been damaged, academic publishers, which have been obliged to withdraw published articles, providers of research funds, whose grants have been used fraudulently, and fellow researchers, who may have been denied the grants and direct funding that were awarded instead to Mr. Stapel.

It's not just direct financial consequences that are impacted by fraud, as the report indicates. It's also the loss of possible opportunities for those who may have been competing honestly for grant money. In addition to the direct monetary costs of grants awarded, many research careers may simply not have been given the opportunity to begin.

We live in a world now where our indiscretions and mistakes are recorded for the world to view. A simple Google search of our name may bring up many pieces of information that we'd prefer the world at large did not see. Around 70% of US employers in 2010 had rejected applicants over what they have found by Googling their names and examining what they have left public on their social networks (Cross-Tab, 2010) – a practice sometimes known as cyber-vetting. A survey by CareerBuilder in 2018 suggested the problem was getting slightly worse, although the prevalence varies a lot by sector. UK employers also engage in this practice, but to a lesser degree (a mere 41% of recruiters have rejected candidates based on online information).

It's important to note too that the information found may not even be accurate – 90% of US recruiters say they are concerned about information accuracy and claim to attempt to corroborate its correctness. In the UK,

80% say they're concerned about accuracy but only 68% claim to make an attempt to corroborate. The prominence and visibility afforded to online information has an impact on what people find about us. A front-page story of a popular newspaper accusing us of something dreadful is likely to be one of the first hits on a search of our name. The much smaller and less prominent retraction by the same newspaper the next day might never be seen.

The Internet does not forget, and the more public our mistakes the more significant the consequences. In the best-case scenario, our postgraduate students might hope to be exonerated, reinstated, and permitted to finish their studies in peace and then move on to promising careers. Unfortunately, the odds are stacked against them. Even if they are found fully innocent by the university, it's entirely possible that future recruiters may never find that out because a disciplinary report on an internal intranet won't have the same visibility as the reporting of Mr. Jack McKracken.

Reputational management is not a matter purely for academics – academia is just an unusual case in that reputation is usually the primary currency in which people deal. Managing one's online footprint is becoming increasingly important. The value of an employer searching for your name and finding your own immaculately designed webpage as the first hit is increasing on a year-by-year basis. It's something that extends outside of our professional lives too – increasingly friends, potential romantic partners (Smith & Duggan, 2013), and those with a financial stake in our future are turning to Google to find out things about us.

When well-managed, our digital footprints may be as much an asset as a liability. A good social media presence with many followers or friends can translate into the impression of access to considerable social capital. Those with large amounts of influence are occasionally courted by organisations looking to wield that influence to mould the perception of their brands on the wider Internet – those that can sell a brand while still appearing authentic demonstrate an increasingly valuable skill (Van Driel & Dumitrica, 2021). A high degree of clout may translate into upgraded flights, access to premium services, or complimentary gifts in the hope that someone will use their social network connections to put in a good word. Increasingly, this status is seen as being so valuable that there has grown a community of people who endorse brands without having deals, in the hope that it inspires other brands to actually pay for endorsement (Lorenz, 2018). Once upon a time, selling out was considered worthy of the greatest social censure. Increasingly though, selling out is seen as an aspirational life goal.

What we're dealing with here though is a more unilateral digital footprint – one that our postgraduate students have little opportunity to manage or edit. It's one likely to dwarf their own attempts to shape a Google search in their interests. As such, it is one that will have genuine consequences for their future employment regardless of the conclusion of the eventual inquiry into their actions.

REMOTE WORKING

One final thing that is worth addressing here is the offhand comment made that the fractious relationship between Blackbriar and his students worsened considerably during the Coronavirus pandemic. This was when working at home became the default model for those in the knowledge-based professions. All of us have our stories about how Coronavirus affected us, and it will likely be a decade or more before the full economic, social, and professional impacts can be assessed. It's already clear that the impact was felt asymmetrically. Some of us – with home offices, perpetual insomnia, and deep-seated introvert tendencies – found it largely compatible with how we'd like work to be arranged as a matter of course. Others – the extroverts (Evans et al., 2022) and those with tiny apartments – found it intolerable. There's no 'one size fits all' conclusion to whether enforced remote working was a boon or a curse. What we can realistically say though is that having had a taste of an alternative, many workers are unwilling to commit to a full-time return to the office (Gibson et al., 2023; Gifford, 2022). Finding a happy balance between work presence and personal convenience is one of the dominant challenges of the post-pandemic economy.

However, the research does suggest that remote working is best when people are already predisposed to work well together. Otherwise, there is a laundry list of problems. People have reported growing isolated (Van Zoonen & Sivunen, 2022) while struggling to deal with a lack of structure (Becker et al., 2022). Some have experienced a sense of career instability (Felstead, 2022), and a difficulty in setting boundaries between home and work (Bellmann & Hubler, 2021). One of the benefits of working in a university is the opportunity for deep, meaningful conversation with thoughtful, intelligent people (Berg & Seeber, 2016). It is much harder to have an impromptu cup of coffee with a colleague when you're all in back-to-back Zoom meetings. Presence of the distractions of home can both inhibit productivity and also exacerbate burnout as it becomes harder to tell where one part of your life ends and another begins.

It is likely some of these factors were at play in the decay of the relationship between Blackbriar and his students. Proximity allows for a lower friction of interaction – scheduling a Zoom call is a more intensive prospect than sticking your head in someone's office and asking a question. Emails can be missed, and tone can be hard to communicate. When everyone assumes good faith from everyone else, these can be minor annoyances and people will work to deal with them. When a relationship is already a problem, the ease with which someone can be avoided can just compound existing problematic dynamics.

However, there is another side to the coin which is that remote working has as many, if not more, upsides – although one must be mindful of the socioeconomic privilege often implied by having convenient, quiet places to work in a home environment. For those with the space to dedicate to their career, home working can be the optimal way to perform meaningful knowledge work. It eliminates the need for commuting (Beno, 2021); enhances autonomy (Aczel et al., 2021); and permits opportunities for a better work-life balance (although that does not manifest equally along gender lines) (Ipsen et al., 2021) while ensuring the optimal blend of working environment to responsibilities. For those employed by a university, working at home is often a core part of how the job is structured. Universities have long understood the value of letting people work closest to their personal optimum when expecting them to be self-directed and in service to their own large self-appointed goals (Jackman et al., 2022).

We cannot then simply point to remote working as the cause, or solution, to any social problems within an institution. It is as much about the remote **workers** as it is about anything else. The roots of the problem in our case study are found in the period predating Coronavirus, but it is likely the enforced shift to remote working was a contributory factor in the further breakdown of the collaboration.

REFERENCES

Aczel, B., Kovacs, M., Van Der Lippe, T., & Szaszi, B. (2021). Researchers working from home: Benefits and challenges. *PLoS One*, *16*(3), e0249127.

Allard, G., Miner, J., Stark, P., & Stevens, A. (2021). *AUTM US licensing activity survey: 2020, a survey report of technology licensing (and related) activity for US academic and nonprofit institutions and technology investment firms.* Association of University Technology Managers.

Andersen, B., & Rossi, F. (2011). Intellectual property governance and knowledge creation in UK universities. *Economics of Innovation and New Technology, 20*(8), 701–725.

Astebro, T., Braunerhjelm, P., & Brostrom, A. (2013). Does academic entrepreneurship pay? *Industrial and Corporate Change, 22*(1), 281–311.

Babor, T. F., & Robaina, K. (2012). Ethical issues related to receiving research funding from the alcohol industry and other commercial. In A. Chapman (Ed.), *Genetic Research on Addiction: Ethics, the Law, and Public Health* (pp. 139–154). Cambridge University Press.

Badea, L., & Mungiu-Pupzan, M. C. (2021). The economic and environmental impact of bitcoin. *IEEE Access, 9*, 48091–48104.

Bastian, H. (2006). 'They would say that, wouldn't they? 'A reader's guide to author and sponsor biases in clinical research. *Journal of the Royal Society of Medicine, 99*(12), 611–614.

Becker, W. J., Belkin, L. Y., Tuskey, S. E., & Conroy, S. A. (2022). Surviving remotely: How job control and loneliness during a forced shift to remote work impacted employee work behaviors and well-being. *Human Resource Management, 61*(4), 449–464.

Bellmann, L., & Hubler, O. (2021). Working from home, job satisfaction and work–life balance–robust or heterogeneous links? *International Journal of Manpower, 42*(3), 424–441.

Beno, M. (2021). Analysis of three potential savings in e-working expenditure. *Frontiers in Sociology, 6*, 675530.

Berg, M., & Seeber, B. K. (2016). *The slow professor: Challenging the culture of speed in the academy.* University of Toronto Press.

Bonaccorsi, A., & Daraio, C. (2007). *Universities and strategic knowledge creation: Specialization and performance in Europe.* Edward Elgar Publishing.

Charles, D., & Conway, C. (2001). *Higher education-business interaction survey.* Centre for Urban and Regional Development Studies.

Cross-Tab. (2010). Online reputation in a connected world.

Evans, A. M., Meyers, M. C., De Calseyde, P. P. V., & Stavrova, O. (2022). Extroversion and conscientiousness predict deteriorating job outcomes during the COVID-19 transition to enforced remote work. *Social Psychological and Personality Science, 13*(3), 781–791.

Felstead, A. (2022). *Remote working: A research overview.* Routledge.

Gibson, C. B., Gilson, L. L., Griffith, T. L., & O'Neill, T. A. (2023). Should employees be required to return to the office? *Organizational Dynamics, 52*(2), 100981.

Gifford, J. (2022). Remote working: Unprecedented increase and a developing research agenda. *Human Resource Development International, 25*(2), 105–113.

Gulbrandsen, M., & Slipersaeter, S. (2007). The third mission and the entrepreneurial university model. In A. Bonaccorsi and C. Daraio (Eds.), *Universities and strategic knowledge creation* (pp. 112–143). Edward Elgar Publishing.

Heron, M. (2011). *The access framework: Reinforcement learning for accessibility and cognitive support for older adults* (Doctoral dissertation, The University of Dundee).

Ipsen, C., van Veldhoven, M., Kirchner, K., & Hansen, J. P. (2021). Six key advantages and disadvantages of working from home in Europe during COVID-19. *International Journal of Environmental Research and Public Health, 18*(4), 1826.

Jackman, P. C., Sanderson, R., Haughey, T. J., Brett, C. E., White, N., Zile, A., Tyrrell, K., & Byrom, N. C. (2022). The impact of the first COVID-19 lockdown in the UK for doctoral and early career researchers. *Higher Education, 84*(4), 705–722.

Krimsky, S., & Rothenberg, L. S. (2001). Conflict of interest policies in science and medical journals: Editorial practices and author disclosures. *Science and Engineering Ethics, 7*(2), 205–218.

Langley, C., & Parkinson, S. (2009). *Science and the corporate agenda: The detrimental effects of commercial influence on science and technology.* Scientists for Global Responsibility (SGR).

Levelt Committee, Drenth Committee, & Noort Committee. (2011). *Interim report regarding the breach of scientific integrity committed by Prof. D.A. Stapel.* Tilburg University.

Lorenz, T. (2018). Rising Instagram stars are posting fake sponsored content. *The Atlantic, 12.*

Lundh, A., Lexchin, J., Mintzes, B., Schroll, J. B., & Bero, L. (2017). Industry sponsorship and research outcome. *Cochrane Database of Systematic Reviews, 2*(2), MR00033.

Macho-Stadler, I., Pérez-Castrillo, D., & Veugelers, R. (2008). Designing contracts for university spin-offs. *Journal of Economics & Management Strategy, 17*(1), 185–218.

Markman, G. D., Siegel, D. S., & Wright, M. (2008). Research and technology commercialization. *Journal of Management Studies, 45*(8), 1401–1423.

Meyer, M., & Tang, P. (2007). Exploring the "value" of academic patents: IP management practices in UK universities and their implications for third-stream indicators. *Scientometrics, 70*(2), 415–440.

Perkmann, M., Salandra, R., Tartari, V., McKelvey, M., & Hughes, A. (2021). Academic engagement: A review of the literature 2011–2019. *Research Policy, 50*(1), 104114.

Perkmann, M., Tartari, V., McKelvey, M., Autio, E., Brostrom, A., D'Este, P., Fini, R., Geuna, A., Grimaldi, R., Hughes, A., et al. (2013). Academic engagement and commercialisation: A review of the literature on university–industry relations. *Research Policy, 42*(2), 423–442.

Pisinger, C., Godtfredsen, N., & Bender, A. M. (2019). A conflict of interest is strongly associated with tobacco industry–favourable results, indicating no harm of e-cigarettes. *Preventive Medicine, 119*, 124–131.

Siegel, D. S., & Wessner, C. (2012). Universities and the success of entrepreneurial ventures: Evidence from the small business innovation research program. *The Journal of Technology Transfer, 37*(4), 404–415.

Smith, A. W., & Duggan, M. (2013). *Online dating & relationship*. Pew Research Center.

Turner, C., & Spilich, G. J. (1997). Research into smoking or nicotine and human cognitive performance: Does the source of funding make a difference? *Addiction, 92*(11), 1423–1426.

Van Driel, L., & Dumitrica, D. (2021). Selling brands while staying "authentic": The professionalization of Instagram influencers. *Convergence, 27*(1), 66–84.

Van Zoonen, W., & Sivunen, A. E. (2022). The impact of remote work and mediated communication frequency on isolation and psychological distress. *European Journal of Work and Organizational Psychology, 31*(4), 610–621.

Washburn, J. (2008). *University, Inc.: The corporate corruption of higher education*. Basic Books.

CHAPTER 5

Multimillion-Pound Consequences for Research Fiddle

NEWSPAPER ARTICLE

An exclusive report by Jack McKracken for the Dunglen Chronicle

The North Sea Algorithmic Exploration (NSAE) project was today the centre of a media storm as the scandal at the University of Dunglen rages outside the boundaries of academia. The multimillion-pound project is supported by a consortium of funding partners including the Scottish Government, the European Union, and 12 of the oil companies involved in drilling for the natural resources within the seas surrounding Scotland. Several universities have been funded to perform research for this project, and Professor John Blackbriar is a significant partner in the work. Senior stakeholders in the project have gone to the press demanding a full public inquiry into how the project has used the Blackbriar data sets and the Blackbriar Algorithms. Initial results from the NSAE had been extremely encouraging, pointing to the potential of additional billions of extractable crude within the region. Lately however, those results have been far less reliable.

Chuma Hassan, spokesperson for the European Funding Council, released a statement today that said, 'We are investigating allegations

DOI: 10.1201/9781003426172-5

of professional misconduct at the NSAE project. We cannot comment at this time but we are fully invested in ensuring the success of this venture. The NSAE project is key to understanding the ways in which difficult to access oil resources can be economically exploited in way that minimises environmental cost. This is a cornerstone of the energy policy of several partner nations'.

'Research is all about trust but verify', an informant within the EFC told the Chronicle today. 'The problem is, NSAE did all kinds of trusting, but not a lot of verifying. There is a lot riding on the success of this project – not just the millions that have been invested by all the funding partners, but billions in what we had thought up to this point were recoverable natural reserves. Really, it's even an issue of world economics – many of us thought we might have less of a reliance of the Middle East as a result of the work. I'm sure you can imagine the consequences if it turns out we've been played for idiots'.

The informant then added, 'I mean, look at what happened with the pandemic and then the war in Ukraine ... I don't know about you, but I've never felt less confident in global supply chains in my life. Local disruption means global impact. None of us work alone'.

While most of the organisations involved in the project have elected to make no formal comment on the allegations, insiders have spoken of the significant worry at the heart of the management boards of the oil companies. A source in ScotOil revealed to us, 'People are panicking – it's not just that people have invested huge amounts of money into the project, but they've also been mentally adding the theoretical yield of the algorithms to the bottom line of the company. This has been going on for years, and it's influenced an awful lot of planning for the next two decades. If you think that there's a good chance you've got a hundred million barrels of oil you can extract beyond what the site managers think, you'd be a fool to ignore that. Rigs that really should have begun their decommissioning have had their lives extended on the grounds that they'll be useful when the new techniques yield results. Now, it's starting to look like they never will – heads are going to roll left, right and centre. Not all of the companies involved will survive this if the allegation turns out to be true'.

The scandal at Dunglen has resulted in an unusual amount of detailed scrutiny being directed towards the oil companies involved in the NSAE project. Senior shareholders are demanding that forensic accountants be allowed access to the full financial records of the company to determine whether mistakes have been made in future planning of exploitation projects. Many are suggesting that they will raise very significant civil claims against the companies, the University, and Professor Blackbriar directly if it turns out that senior managers have been counting this revenue before it's even been collected.

It is not clear at this time how much of the tainted Blackbriar data sets have been used within the NSAE project, but it seems likely that they were a significant reason why early simulations and modelling projected significant extractable reserves. 'We ran the details of one field that was completely tapped out through the Blackbriar Algorithm', our ScotOil insider said, 'And found out that there was maybe £250m of extra revenue there. That's the kind of money that can cause otherwise level headed people to make serious mistakes'

Vanessa Haynes, research director of the NSAE project, said, 'While these are serious allegations, it's important to realise that they have yet to be substantiated. However, even if they should prove to be valid the impact on NSAE is likely to be minimal. Our systems and projections are based on several externally validated data sets, and while Professor Blackbriar is a key player in our academic partnership he is not our only source of data. The Blackbriar Algorithm is a significant tool in our analytical routines, but if it turns out to be unreliable we will retire it. We will rely instead upon the other indicators we use to identify potentially high-profit yields from otherwise uneconomic oil and gas fields'.

'All fine and well', our ScotOil insider says. 'But while that may mean future results are more accurate, we've got a lot riding on the past results being true. There are a lot of jobs on the line here'.

ACADEMIA AND THE CORPORATION

Within the article 'Multimillion-Pound Consequences for Research Fiddle', the Scandal in Academia discusses the relationship between the University of Dunglen and its industrial partners in the North Sea. We find out that there is a great, and growing, worry that the NSAE project may have based much of its future planning on research data that has become suspect. There is talk of lawsuits being levelled at everyone involved. It's an awkward situation and one with potential professional and personal consequences for a large number of partners. We've already discussed some of this, but we'll find as we go along certain themes keep emerging as we explore the scenario from different angles.

The tensions between industrial and educational research policies and philosophies are a long-standing, ongoing issue. Ethical standards, levels of confidence, and requirements for corroboration all vary enormously within academic disciplines; within research practitioners within disciplines; and within the interfaces between university departments and faculties. Expectation of authorship on papers may not be shared by all and the level of assumed contribution before co-authorship is conceded may create tensions. Within academia itself, collaboration can be politically vexatious. When industrial collaborations are incorporated into this, the problems magnify because not only do large industrial partners suffer from the same issues, they also approach the matter from an often completely different perspective. Thus, in addition to internecine conflict within organisations, there is an external conflict at the interfaces between groups. No matter the will to collaborate or the mutual respect that may be found on both sides, a lack of common ground can create difficulties, and these in turn can lead to violations of cultural and professional norms when the work is to be published. Issues of commercial sensitivity, for example, are anathema to principles of free academic disclosure.

Within academia, the differentiation between 'research' and 'teaching' is usually well observed – as a matter of course, academic research is not conducted directly on students unless it is educational in nature and students explicitly opt in with no coercion. Research projects are clearly demarked, at least in theory, and have formal start and termination points. The funding for academic research is most often provided by external bodies and this places an emphasis on clearly delimiting activities for the purposes of economic auditing and accountability. Most university governing bodies demand strict adherence to ethical codes of conduct, and research which involves working with real people at any time will usually be scrutinised for its conformance with overall university codes of practices. It is usually possible then, within a university, to point at a body of work and say, 'This forms part of this distinct research project, which was led by this individual, and funded by this external body in this particular way'.

Within industry, these distinctions can become blurred – there is a fine line between 'improving a process' and 'testing a hypothesis'. Consider for example the widely publicised case of Facebook's 'experiments' over the emotional impact of manipulating the news feeds of their users (Kramer et al., 2014). Much heat was generated as a result of Facebook, in collaboration with university academics, publishing some conclusions they had drawn from what was essentially a large-scale experiment on human subjects through modification of the Facebook news feed. Commentators variously described the research as unethical, noble, deceptive or as a violation of privacy (Hallinan et al., 2020; Kramer et al., 2014). Leaving aside the efficacy or value of the research itself – which is disputed – what is most clear about this incident is the differing expectation of what research actually involves within an industrial context. Within academia, ethics forms would be submitted, funding obtained, proposals scrutinised before being approved, and the results would be submitted to a journal for full peer review, followed by an affirmation that sector norms for ethical conduct had been observed. Within Facebook, and other large-scale industry organisations, it may be a common part of day-to-day practice to engage in repeated and regular A/B testing on cohorts to improve, for some given value, the interaction experience for all.

That testing may involve adjusting the load order of dynamic web elements, changing the algorithms used to retrieve information, or altering the precedence given to information as it is presented. The agility that industrial organisations can muster for this in many ways determines their ability to keep pace with their competitors, especially in fast moving fields such as computing. For some time, the mantra at Facebook was 'Move fast, and break things'. That's not reflective of an attitude focused on mature deliberation of cause, effect, and consequence.

For most corporations, the results of such testing are usually kept internal and employees are bound by their existing employment contracts and codes of conduct. The tension arises when this research transgresses this 'invisible' boundary and interfaces with both academia and the public. The Facebook research was published in the Proceedings of the National Academy of Science of the United States of America (PNAS) as a collaboration between Facebook, the University of California, and Cornell University. The PNAS has as one of their submission criteria a requirement that research is conducted according to the Declaration of Helsinki. This is where the offence lies, because the research does not meet that criteria. Had the research never been published however, there would have likely been no outcry. Within large organisations, such 'research' is largely just a part and parcel of an ongoing adjustment and refactoring of internal systems.

For academics, access to large and restricted data sets can be seductive – the original paper (Kramer et al., 2014) had a sample size of just over 689,000 users. It simply would not be possible for an academic partner, acting in isolation, to recruit so many people to a trial. For industrial partners, the credibility of a respected academic collaborator can burnish up results, as well as ensure that they are presented, analysed, and contextualised according to the rigorous expectations of primary research publications. Drafting research for academic outlets is a specialised skill and requires familiarity with the linguistic conventions not just of the discipline, but the editorial policies of the journal or conference and the wider sector beyond.

Consent for participation in such internal research by companies such as Facebook and Google is usually permitted as part of a blanket acceptance of terms and conditions on the part of the users. In the case of the Facebook study, there is some evidence to suggest that consent for research was added to the user agreement after the research had been conducted but in most cases a blanket exemption is in place that covers the service provider for a wide range of activities. This in itself is a shallow defence, given how few people read the terms and conditions and how explicitly impenetrable they are often made to be (Luger et al., 2013). Core to the objection that many have had to the research is that while it may have received consent in its simplest, most shallow form, it doesn't meet the criteria for informed consent which has gradually become the academic consensus since the Second World War (Vollmann & Winau, 1996). Industrial bodies, such as Facebook, are rarely bound by formal codes of conduct or even institutional review boards (IRBs) that sanction studies. Even for those involving human participants. Universities, on the other hand, tend to have a more institutionally rigorous infrastructure for reviewing the ethical implications of research. That said, the standards for this do vary considerably from country to country.

However, the tensions of corporate influence in academia extend beyond simple differences in expectations of ethical consent. One of the primary ways in which industry can partner with academia is through the route of funding or sponsorship. We discussed the impact of commercial interests in our last chapter, but we also have to consider the powerful influences that corporate money can have on academic freedom. While there are real-world limits on just how far academic freedom stretches, it is in general considered to be a principle worth defending. Support for the principle though is tempered. There exist tensions – for example, advocates of Intelligent Design as a scientific principle have sometimes claimed to find it difficult to obtain advancement in their institutions. It is hard though to unpick in such cases the ratio of academic suppression as compared to the self-correcting nature of the scientific process. Within the United Kingdom, the Education Reform Act of 1988 codified academic freedom as a guiding principle for higher education institutes. Even though academics are under legal protection that does not prevent industrial partners from leveraging the power differential implied by funding to suppress results or to encourage undue prominence being given to minority views that coincide with their economic interests, corporate interests can result in delays on publication, or harmful secrecy clauses on results obtained (Kramer et al., 2014). They can result in papers or reports with written conclusions that are contrary to an analysis of the data, because it is the introduction and conclusion to which many popular press outlets will refer. They can result in academics acting in part as the Public Relations arm for corporate influence within specialist publications. Consider for example the role played by Stossel in Brody (2008). The support of an academic, or their institutional brand, can bolster the reputation of an organisation. Such bolstering may be worth the financial outlay if it can be leveraged effectively.

Fairly or unfairly, the source of funding for research has become as important an element of full disclosure as data sets and methods. There is an element of backlash against this expectation, alleging it has become a tool through which vested interests can undermine unpopular research and that this in turn hinders progress (Stossel & Stell, 2011). Stossel is quoted in Jupiter and Burke (2013) as saying the following:

> Disclosure policies are no longer a way to honour the sponsor of a study, but rather they have been turned into a type of confession. In practise, disclosures are being used by the media to embarrass people. We have gone from bad to worse. We have immense regulatory issues and massive confessions where we disclose our relationships to industry and those are used to initiate a variety of inhibitions of freedom of speech, freedom of association, and rewards for excellence.

However, a critic may argue that only those things which are commonly considered to be shameful can be used to embarrass. If disclosing a relationship with a funder is shameful, then it is symptomatic of the often compromising relationship between industry and academia. The now infamous Paxil Study 329 (Jureidini et al., 2008; Washburn, 2011) liberally selected from those data points most likely to show a positive result of the medication under review. In this respect, the principal investigator for this shows a very similar pattern of work to what is alleged of our own Professor Blackbriar. He too, it is claimed, has adjusted the data of his own research in part to assuage the powerful corporate interests who have been funding his work.

MENTAL BIASES

In our case study though we are not talking about how corporate sponsorship may violate the principles of research, but instead about how research may undermine the work of industry. Here we have an experimental algorithm which has been incorporated into a large-scale industrial project, and it's not giving the results people are hoping for. A lot is riding on the success of the project, and that success is highly tied up in the question of whether or not the Blackbriar Algorithm works at all.

It's easy for an external party to look at some situations that occur and think 'well, the real solution is to not get into that situation in the first place'. This is superficially compelling, but ignores the vast array of factors that result in bad decisions being made.

In teaching computer ethics, the first task we have as educators is to disabuse students of the idea that it's all 'just common sense' (Heron & Belford, 2015). We must counter the idea that we as individuals are immune to the psychological and social factors that influence others. In fact, the belief that we are somehow calmer and more rational than others is in itself a demonstration of the cognitive biases under which our minds labour. This particular bias is called illusory superiority (Hoorens, 1995). Most of us believe, regardless of the illogicality of the statistical implication – that we are above average. Most of us tend to believe that what impacts on others will not impact on us quite so intensely.

The human mind is a remarkable tool, honed by evolution to a fine point. However, it is also a product of our historical context, and as such it contains not only the cognitive architecture that we need for human society as it is now, but as it was hundreds of thousands of years ago. The human subconscious is also a tremendous filter, protecting our conscious mind from the vast amount of information our senses pull in on a second-by-second basis. As part of that apparatus, there is a need for our mind to protect against cognitive overload. Often, this is done through cognitive shortcuts that allow us to quickly arrive at a judgement that is 'good enough' for most purposes. These lead to cognitive biases which systematically influence the way in which we think about the world. The collection of these identified biases is considerable and many of them are relevant to the issue of why people make bad decisions.

Perhaps most germane to this specific example is the Sunk Cost effect, or what is often known as the Concorde Fallacy (Arkes & Blumer, 1985).

When Concorde was being built, the British and French governments continued development long beyond the point where there was a reasonable expectation of economic return on the project (Arkes & Ayton, 1999). So much time, effort, and political capital had been invested in the project that it was inconceivable to simply stop the work. However, economic theory argues that when deciding on a future investment, sunk costs should be entirely discounted. They have been spent regardless of the success or failure of the project and thus should not feature into future decision-making. This is a difficult lesson to internalise for most people, and this difficulty leads to what is colloquially known as 'throwing good money after bad'. After investing so much money into Blackbriar and his work, it may have been the case that his funding partners simply decided they had invested too much already to abandon the project.

This though is predicated on an assumed understanding on the part of the research partners – it presumes that they understood the algorithm as it was presented was a bad investment. However, we have other biases that stop us being able to make that kind of rational assessment when it comes to complex sets of information. Consider for example the confirmation bias – the cognitive shortcut that leads us to preferentially seek out (or consciously notice) information that supports our existing beliefs (Keates & Clarkson, 2003). If we believe that the algorithm works, we're more likely to 'weight' evidence that it works in our mind – even if that evidence is not as strong or as common as evidence that it doesn't. Likewise, the hindsight bias (Arkes et al., 1981) is relevant. This leads us to a kind of mental revisionism in which we believe we had assumed an event was going to happen the way it did all along even if we had no way of predicting. The hindsight bias can lead to memory distortions in which we not only believed we were right but remembered actively predicting that the result would occur.

There are a lot of funding partners in the NSAE, so we also have to consider socially contextual cognitive biases. Consider the bandwagon effect – that the more we encounter people who believe a thing, the more likely we are to believe that thing ourselves regardless of the underlying evidence. That in turn leads to a sampling bias where we are mistakenly led to believe that those around us are representative of a group at large, even though they may be an unusually skewed cluster. This then can lead to groupthink (Janis, 2008), in which presumed consensus acts as a barrier to exploration of risks, pitfalls, and counter arguments. Or consider the Status Quo bias (Kahneman et al., 1991), where we are more likely to accept a situation as it is presented rather than attempt to change it.

The latter bias has been manipulated to startling effect in 'opt-out' versus 'opt-in' initiatives. If participants are asked to opt in to an organ donor register, donor rates hover about 15%. If people are asked to opt out, donor rates rise to around 80% (Davidai et al., 2012; Johnson & Goldstein, 2003). Most people simply will not tick the 'opt-out' box, choosing instead to accept the default with which they were presented. The power of this effect has been noticed; you may have seen yourself many websites requiring that you opt out of receiving their email notifications upon registration, as opposed to opting in.

Even leaving aside these cognitive biases, we can't discount the simple placement or emphasis of information and the role it plays in prompting a decision. Consider for example the two related concepts of anchoring (Tversky & Kahneman, 1992) and framing (Tversky & Kahneman, 1981). Anchoring refers to the technique of setting early expectations in a comparison, and framing refers to the selective placement or information as it is presented. Presenting relatively large default sums anchors our value judgement – when we decide to pay a lesser sum, it is a larger lesser sum than it would have been if there were no anchoring offered at all. In this way, we are skewed towards paying more than we otherwise may have.

Framing works by offering multiple options at once, with one being contextually much better than the others. Ariely (2008) discusses one such example of this in relation to the subscription options for the Economist:

> I read these offers one at a time. The first offer – the Internet subscription for $59 seemed reasonable. The second option – the $125 print subscription seemed a bit expensive, but still reasonable.
>
> But then I read the third option: a print **and** Internet subscription for $125. I read it twice before my eye ran back to the previous options. Who would want to buy the print option alone, I wondered, when both the Internet and the print subscriptions were offered for the same price? Now, the print-only option may have been a typographical error, but I suspect that the clever people at the Economist's London offices (and they are clever-and quite mischievous in a British sort of way) were actually manipulating me. I am pretty certain that they wanted me to skip the Internet-only option (which they assumed would be my choice, since I was reading the advertisement on the Web) and jump to the more expensive option: Internet and print.

In this case, the option is framed in such a way as to skew the choices people make – they're more likely to go for digital + print even if they just want a digital copy purely because that deal seems like it's so much better than the print option alone. With the print+digital option, it appears as if digital access is a free bonus of having purchased print access. Such techniques are used often in retail to skew people towards a 'mid-range' option in circumstances where they may have otherwise purchased a cheap option. Often, this kind of 'nudging' is entirely incidental, but it can be used to subtly, and powerfully, change the way in which our minds analyse the information in front of us. It is easy to conceive of a meeting in which an academic pitches four projects – one extremely expensive, one merely very expensive, one cheap, and one that is a more middling cost. If each was presented individually, the cost may have been prohibitive for all but the cheapest option. If presented collectively, or framed, the moderate option becomes more attractive because it is seen in comparison to two expensive options, rather than assessed on its own merits.

There are many more of these biases that are relevant to the issue of sensible decision-making – there's the Gambler's Fallacy (Tversky & Kahneman, 1971) or the IKEA effect (Norton et al., 2012) or the Optimism Bias (Stossel & Stell, 2011) or the Experimenter's Bias (Sackett, 1979) and many more than we can hope to even touch on in this short section. However, the key point here is not that any of these biases **were** in play, but that they **could** very easily have been in play. Any one of them would have impacted on cool, rational decision-making – often without the people involved having any idea that there was anything wrong with their thought processes. We often assume that our decisions are the result of the calm, sober application of rational analysis. To assume such is to ignore the gaping holes in our minds through which conscious or unconscious manipulation can enter our thinking. It is easy to judge based on hindsight but in order to understand what may happen in the future, we need to be mindful of the limitations of our mental architecture. We need to be aware that, if placed in a situation directly, we are likely to be subject to the same powerful cognitive forces as others.

THE GLOBAL ECONOMY

One thing that is mentioned by the anonymous informant in this article is the supply-chain implications associated with global events. At the time of writing, in 2023, we are still reeling as a world economy. We're seeing record levels of inflation across the board but especially in regard to electricity and food prices. It was even worse in 2022, with July of that year representing the highest US inflation since 1981 – a 40-year high. The situation was similar in the European Union, the UK, and in Russia. Double-digit inflation increases – within robust economies which typically viewed 2% inflation as the acceptable target – created an immediate and intense cost of living crisis. Countries lacking a robust and diversified economy that could absorb shocks suffered even worse. The causes were multivariate – our informant in the article refers to the pandemic and the war in Ukraine as contributory factors, but there are many others. The United Nations in 2022 published an analysis of the data, suggesting that the real-world impact was to push over 50 million additional people into 'extreme poverty' which is defined as having $1.90 a day or less on which to live. Over 20 million more found themselves with $3.20 or less to cover their needs. However, these macroeconomic consequences seem to be at the wrong level to concern those working with the Blackbriar Algorithm, surely?

It is certainly true that nobody here is at risk of ending up in extreme poverty, but we need to take others into account. One of the biggest impact factors on global electricity prices has been volatility in the oil market. In 2019, Taghizadeh-Hesary et al. argued that around 64% of variance in food prices over the period 2000 to 2016 could be explained by a change in oil prices. Sanctions between the West and Moscow in 2022 have created the perfect conditions for oil price volatility (Appiah-Otoo, 2023), and given the scale of the expectations we're dealing with in our case study (billions of dollars) it seems likely that this uncertainty would only be exacerbated by our scandal as it continues to grow in the attention it receives. It may seem a tenuous link, but we can see a clear path of rhetoric that takes us from 'questionable data used in a Scottish university research project' to 'people dying of starvation'. The common adage 'Think globally, act locally' is a double-edged sword. Sometimes the consequences of decisions are far beyond our reasonable expectations.

Realistically, we'd also be likely to see the effect of our scandal manifest in complications with the Just-In-Time (JIT) economy that crashed so spectacularly during the Coronavirus pandemic. A moderate rise in oil prices might be enough to put a trucking firm out of business, or for the contributory effect on global inflation to require an agri-farm to shed employees and slash production forecasts. Modern technological logistics tools – many of them based on software and predictive artificial intelligence – have revolutionised supply chains across the world. No longer does a factory need to maintain an expensive warehouse full of stockpiled parts. Local inventory needs can be minimised by holding only what is immediately required for short-term production. The expectation within these systems is that other parts of the supply chain can ensure reliable, regular, and predictable delivery when that local stock is depleted. By building long-term contracts with trustworthy suppliers – those with a history of delivering on-time and on-budget – efficiencies in the supply chain can boost profitability. When it works well, it can be revolutionary.

However, it is a fragile and risky proposition because the longer the JIT chain is, the more vulnerable it is to shock. You may have a store selling tables, which relies on a factory that assembles those tables, which in turn relies on a workshop that creates the parts of the table, who themselves rely on a shipping company to transport raw wood, who relies on a logistics firm in another country to get the wood to a port, who in turn relies on a logging camp. That logging camp itself may be relying on a JIT chain for tools and people. All it takes is a few unexpected accidents with a chainsaw for you, the person who just spent thousands of pounds on a prestige table of Norwegian spruce, to be told the item you ordered won't be with you for a couple of additional months. Brexit has been convincingly linked to many of the inflationary pressures within the UK, as additional complications in the supply chain have knock-on effects. That's fair, but realistically all it takes is a snap of cold weather in a foreign country for your local supermarket to find it difficult to source the fresh fruit and vegetables you may be expecting. And even for those items available the price can shoot up as supply and demand issues start to take hold. Supermarkets can only ever hold a small amount of fresh produce at any time, and when supermarket shelves run bare human nature starts to take over. Despite repeated calls for consumer calm from supermarkets, panic over the availability of common items resulted in global stockpiling in the early days of the pandemic (Micalizzi et al., 2021). Largely this was unnecessary – nobody needed to buy six-month worth of toilet paper regardless of how much supply-chain volatility might have impacted on their diet. Supermarkets can usually source around 60-day worth of non-perishable items even in the face of supply-chain problems. A quote often attributed to Vladimir Lenin is that 'Any society is only three meals away from chaos', and the **perception** of scarcity can have bigger societal impact than the scarcity itself. It can create a self-fulfilling prophecy – we believe these goods are scarce. We will buy them in unanticipated bulk. The supply chains did not anticipate this surge in demand and failed to resupply in time or in bulk.

Therefore, the goods become scarce. Therefore, our belief was correct. Therefore, buy all the toilet paper you can.

It's unlikely that the question marks over the Blackbriar Algorithm's accuracy will destabilise economies. However, it's important to appreciate just how interconnected the modern world is – there is a direct line to be drawn from factory shutdowns in China to the inflated cost of cars in the UK. Modern software systems have allowed for unprecedented levels of integration, but all software encodes human assumptions and sometimes those assumptions don't take things like pandemics and wars into account.

REFERENCES

Appiah-Otoo, I. (2023). Russia-Ukraine war and us oil prices. *Energy Research Letters*, *4*(1), 1–5.

Ariely, D. (2008). *Predictably irrational*. Harper Audio.

Arkes, H. R., & Ayton, P. (1999). The sunk cost and Concorde effects: Are humans less rational than lower animals? *Psychological Bulletin*, *125*(5), 591.

Arkes, H. R., & Blumer, C. (1985). The psychology of sunk cost. *Organizational Behavior and Human Decision Processes*, *35*(1), 124–140.

Arkes, H. R., Wortmann, R. L., Saville, P. D., & Harkness, A. R. (1981). Hindsight bias among physicians weighing the likelihood of diagnoses. *Journal of Applied Psychology*, *66*(2), 252.

Brody, H. (2008). A reply to Thomas Stossel on the AMA-CEJA draft report. *The Medscape Journal of Medicine*, *10*(7), 154.

Davidai, S., Gilovich, T., & Ross, L. D. (2012). The meaning of default options for potential organ donors. *Proceedings of the National Academy of Sciences USA*, *109*(38), 15201–15205.

Hallinan, B., Brubaker, J. R., & Fiesler, C. (2020). Unexpected expectations: Public reaction to the Facebook emotional contagion study. *New Media & Society*, *22*(6), 1076–1094.

Heron, M. J., & Belford, P. H. (2015). A practitioner reflection on teaching computer ethics with case studies and psychology. *Brookes eJournal of Learning and Teaching*, *7*(1).

Hoorens, V. (1995). Self-favoring biases, self-presentation, and the self-other asymmetry in social comparison. *Journal of Personality*, *63*(4), 793–817.

Janis, I. L. (2008). Groupthink. *IEEE Engineering Management Review*, *36*(1), 36.

Johnson, E. J., & Goldstein, D. (2003). Do defaults save lives? *Science*, *302*(5649), 1338–1339.

Jupiter, J., & Burke, D. (2013). Scott's parabola and the rise of the medical-industrial complex. *Hand*, *8*(3), 249–252.

Jureidini, J. N., McHenry, L. B., & Mansfield, P. R. (2008). Clinical trials and drug promotion: Selective reporting of study 329. *International Journal of Risk & Safety in Medicine*, *20*(1–2), 73–81.

Kahneman, D., Knetsch, J. L., & Thaler, R. H. (1991). Anomalies: The endowment effect, loss aversion, and status quo bias. *Journal of Economic Perspectives*, *5*(1), 193–206.

Keates, S., & Clarkson, J. (2003). Countering design exclusion. In *Inclusive design* (pp. 438–453). Springer.

Kramer, A. D., Guillory, J. E., & Hancock, J. T. (2014). Experimental evidence of massive-scale emotional contagion through social networks. *Proceedings of the National Academy of Sciences USA*, *111*(24), 8788–8790.

Luger, E., Moran, S., & Rodden, T. (2013). Consent for all: Revealing the hidden complexity of terms and conditions. In *Proceedings of the SIGCHI conference on Human factors in computing systems (CHI '13)* (pp. 2687–2696). Association for Computing Machinery, New York, NY. https://doi.org/10.1145/2470654.2481371.

Micalizzi, L., Zambrotta, N. S., & Bernstein, M. H. (2021). Stockpiling in the time of COVID-19. *British Journal of Health Psychology*, *26*(2), 535–543.

Norton, M. I., Mochon, D., & Ariely, D. (2012). The IKEA effect: When labor leads to love. *Journal of Consumer Psychology*, *22*(3), 453–460.

Sackett, D. L. (1979). Bias in analytic research. In *The case-control study consensus and controversy* (pp. 51–63). Elsevier.

Stossel, T. P., & Stell, L. K. (2011). Time to 'walk the walk' about industry ties to enhance health. *Nature Medicine*, *17*(4), 437–438.

Taghizadeh-Hesary, F., Rasoulinezhad, E., & Yoshino, N. (2019). Energy and food security: Linkages through price volatility. *Energy Policy*, *128*, 796–806.

Tversky, A., & Kahneman, D. (1971). Belief in the law of small numbers. *Psychological Bulletin*, *76*(2), 105.

Tversky, A., & Kahneman, D. (1981). The framing of decisions and the psychology of choice. *Science*, *211*(4481), 453–458.

Tversky, A., & Kahneman, D. (1992). Advances in prospect theory: Cumulative representation of uncertainty. *Journal of Risk and Uncertainty*, *5*(4), 297–323.

Vollmann, J., & Winau, R. (1996). Informed consent in human experimentation before the Nuremberg code. *BMJ*, *313*(7070), 1445–1447.

Washburn, J. (2011). Academic freedom and the corporate university. *Academe*, *97*(1), 8–13.

CHAPTER 6

Students Speak Out

NEWSPAPER ARTICLE

An exclusive report by Jack McKracken for the Dunglen Chronicle

The two doctoral students at the centre of what is becoming known as the BrokenBriar Affair have spoken exclusively to the Dunglen Chronicle today in an effort to make clear their side of the story. The students were suspended several weeks ago as a consequence of their work being associated with the increasingly questionable data sets that were curated by Professor John Blackbriar. Mrs Sharon McAlpine and Mr James Duncan found themselves at the centre of the controversy when it was revealed that papers they had co-authored with Blackbriar had been based on suspect data.

'John is an excellent researcher, but a terrible supervisor', said Duncan. 'He doesn't have a lot of time for his postgrads or even his postdocs – he's all about publishing, publishing, publishing. He doesn't even do any teaching these days – he leaves that to us. Our job is to generate papers and ascribe him authorship so as to ensure his academic profile continues to grow'. Duncan then added, 'After the pandemic hit, we didn't even see him in person. A lot of us felt that out of sight was out of mind'.

McAlpine adds, 'You have to understand what it's like for postgrads like James and I – being linked to Professor Blackbriar is a huge deal. He's a titan in the field, and his word can make or break your future. A recommendation from him, at least until recently, was all but a guarantee of a great job once we'd graduated. If we wanted to stay in academia, his name on a grant application was the closest thing to sure-fire

DOI: 10.1201/9781003426172-6

postdoctoral funding. However, if you didn't end up on his good side you were looking at your future being spent across three or four short-term contracts as a kind of nomadic academic. I'd be in my mid-thirties before I had anything approaching career stability – if I ever achieved it. The way my burden of domestic work skyrocketed during the pandemic meant that I fell behind relative to my mostly male colleagues who didn't have the same unpaid labour expectation upon them'.

Duncan and McAlpine describe their relationship to their supervisor as something akin to semi-willing serfs. They taught his classes, did his marking, and handled his less important public outreach responsibilities. They did lab experiments for him and crunched numbers for statistical analysis. In return, he gave them preferential access to his data sets and would vouch for them in future engagements. His supervision, we are told, was largely non-existent.

Duncan laughs, saying, 'One time it was pretty clear that he'd asked an AI chatbot to summarise one of our reports, and then passed off its output as his own thoughts. Once he said that he'd take one of his team to an important research retreat based on our performance at a "research exercise". After months of asking which of us were going, he eventually said that none of us had made the grade. I heard later from his personal assistant that he'd never even looked at them'.

McAlpine elaborates, saying, 'Don't get me wrong, we weren't just "left loose in the building". We each had regular meetings with other members of staff to discuss our progress, and we had various transfer meetings and internal research discussions as we went through our doctoral studies. But it's hard, you know, to say anything about what's going on. John was quite ruthless in his collaborations – he knew his reputation was valuable to others as well as to himself, and he used that as leverage all the time. He also had a short temper and people could easily end up on his bad side for no real reason. With that in mind, we were in the position of needing to balance short-term wants against long-term needs. Life was full of promise if you were one of his current favourites. Life looked pretty bleak if you weren't, and you could go from one group to the other overnight without saying a word to him'.

When asked about their collaboration on the writing of papers, Duncan says, 'Sharon and I worked together on the construction of the first paper. All John did was read it over and make sure that we'd done the calculations correctly. Then he insisted on first author credit for it. That was okay – we knew that people would read and cite a paper he was named on, and that they probably wouldn't pay any attention to unknowns like us. The second paper was almost exclusively written by Sharon – with some judicious assistance from an

AI Chatbot in the discussion section. She was kind enough to put me as third author since I'd helped with the planning and reviewing of the paper. She fought John against his demand she accepted second author position and eventually submitted it with her name first. He conceded because it didn't use any of his data directly. Really, he didn't contribute anything to it but as supervisor and source of our funding he felt he had authorship rights to everything we publish'.

McAlpine responds to allegation of incompetence or fraud: 'The allegations are total bullshit. The papers may be based on fraudulent data, but we don't know that for sure yet. But if they are, it's not like it's our fault – we didn't have access to the full data sets, all we could do was make sure that what we published was accurate to the best of our knowledge and we did that. We ran those models dozens of times, redid the maths, checked our conclusions rigorously. What we published was absolutely correct going by the data going in. It's not our fault that the data is questionable. Garbage in, garbage out as they say – but we didn't have any opportunity to tell that it was garbage going in'.

Duncan adds, 'You have to bear in mind too the pressures we were under. It was just expected we'd work from 8am to 8pm every workday, and if you didn't come in on weekends too you'd be questioned on it. Blackbriar used to say "If you're not willing to put in the work I can whistle and have ten people here tomorrow who are". We all knew that was true, so we all worked "voluntary" seventy- or eighty-hour weeks every week. We were doing his teaching, his marking, even some of his internal administration. We were his lab technicians as well as his PhD students and we spent days running tests on whatever he wanted to check out even if it wasn't remotely relevant to our work. And on top of all that, we heard the constant mantra "publish or perish". Even if we did have access to all the data and even if could validate it, we just wouldn't have had the time. It's petabytes of information. For my part, it's even more complicated than that – all the data was contained within a bespoke application written by one of Blackbriar's previous research students. I'm legally blind, and the software didn't work with a screen-reader. I couldn't even get access to the data I was permitted – I needed Sharon to do that for me. She went in, grabbed the results, and put them into a spreadsheet I could use. But that was a long, laborious process and I couldn't ask her to do it every time I needed data. We experimented with building some AI tools into the software to ease with the extraction of what I needed, but the results were never fully reliable. And in cases of precise calculation with complex values, "mostly correct" is as bad as "never correct"'.

McAlpine expanded on her problems with the process. 'I was a new mother, and I can't stress to you how much Blackbriar hated what that meant. He expected us to be "always available", which meant that if he had an idea at 7pm you'd better be prepared to work on it overnight and report progress the next morning. That's just not possible when you have a baby. Or if you want a healthy life. Then, of course, the pandemic hit and suddenly I found that not only was I working at home with almost no support from my husband but I was expected to find room to work in a two-bedroom flat with two people working permanently at home. He was the one with the "career", so he got the spare room and I got the kitchen table. When housework needed to be done, he was the one in high-powered meetings and couldn't take time out. I was "just a student" and thus I could be a full-time home-maker as well as a researcher. At least, that was his theory'.

When asked if the 'his' in that comment related to Blackbriar or her husband, McAlpine simply answered 'yes'. When asked if their situation was known beyond their research group, both are emphatic that it was an 'open secret' and 'not even far out of the average during the pandemic'.

'Everyone knew', says McAlpine, 'But nobody made a big thing of it. I know that Professor Tumblewood had shouted loudly to the various members of the Professoriate that grant generation was vital and we had to get maximum value out of "research resources". He didn't mean lab equipment. He meant postgrads and postdocs. That was just the culture – lots of people criticised it, but nobody did anything about it'.

'It became the norm', adds Duncan. 'As it started to take hold in the "elite" research groups, the smaller or less important groups adopted a similar approach. After a while, it just became "the way we do things". Work-life balance was something that you earned, not something you were allowed to have walking in the door as a new postgrad'.

We approached both Professor Blackbriar and the University of Dunglen about the remarks from their suspended students. Professor Blackbriar was unavailable for comment, and the university issued the following statement:

`We are investigating Mrs. McAlpine and Mr. Duncan for charges of academic fraud. We cannot comment on the specifics of their allegations against the culture of research within the university while these investigations are ongoing. However, we have an unparalleled reputation for our research conduct, and the postgraduate and postdoctoral positions within our most highly respected research centres have application over-subscriptions in the order of eighty to one. People very much want to be a part of our research community. Whether there is any truth in the suggestion that we offer an unpleasant working environment for researchers should be assessed in light of that fact.`

WORKPLACE RELATIONS

Within the 'Students Speak Out article', we see the first stirrings of the workplace issues that will become important themes. Our postgraduate students, Sharon and James, have been suspended for their presumed roles in the alleged academic misconduct. They have told their story to the newspaper in an attempt to get their views heard. What they tell is a tale of academic serfdom and the hope of future preferment because of their relationship with Blackbriar. What they also do is open our eyes to the way in which modern academia sometimes utilises transient resources such as postgraduate and postdoctoral researchers. There are parallels here too to the way other industries treat young professionals – burn them out and then change them like lightbulbs.

There is a growing body of what has become known as 'quit-lit' emerging in the semipopular educational press. This term broadly encompasses a range of revelatory blog posts, education periodical editorials, and social media updates. As a general theme, these revelations cover postgraduate students, postdoctoral researchers, and even full-time faculty members. Those who have been driven to publicly quit their positions as a result of administrative pressures, job insecurity, or career ennui. The stories are not just from those who have failed to find success within modern academic institutional structures, but also those who have found such success and discarded it regardless. Full quantitative figures on the trend are hard to uncover, as unpicking these incidents out of larger employment trends is a complex task. It is hard to say whether the trend to publish 'I'm leaving academia' literature reflects an increase in dissatisfaction or simply a decrease in discretion. However, within the body of quit-lit we see many views of a dehumanising system of employment and promotion, and a research process which prizes funding and quantification over longer term scholarship. Obtaining a permanent position as an academic is difficult. There are some 30 or 40 PhD graduates being produced for every single academic vacancy, with positions at high-profile institutions sometimes receiving hundreds of applications per job. Under such circumstances, it is only natural that those with transient working contracts will look for whatever advantage they can find in their collaborations. Getting noticed as a new researcher too is challenging – with little track record of individual accomplishment, grant funding is difficult to obtain. Temporary research contracts usually relegate a researcher to second or third authorship. Attaching one's name to a prominent researcher in the field can be a useful way to gain some notice. A certain amount of professional discomfiture might be expected and accepted as the cost of doing business. A kind of 'competent-by-association' impression can be generated by your name being attached, via co-authorship, to the prominent publications of an academic luminary.

The growing trend of quantification of research exacerbates this issue – assessment exercises such as the UK's Research Excellence Framework place considerable weight on algorithmic analyses of research output, such as citation counts and H-Indices (Martin, 2011). Automated tools such as Google Scholar index scholarly publications across much of the internet and produce crawled citation lists. These are an easily checked resource for both researchers and those looking to employ them. Other databases exist too, with varying degrees of prestige. For some disciplines, all that matters is how many publications you have in Scopus, for others it's the Web of Science. We've spoken before about fractures at the interface of disciplines, and this is one of the ways in which it manifests in academia. Not all databases capture all publications – they're all skewed towards one discipline or practice or another.

Anything quantifiable can almost always also be gamed, and the H-Index is no exception. If you haven't encountered this term before, it is described as 'The value N, such as N publications from a researcher have achieved N citations'. As in, if you have five papers with five citations, you have an H-Index of five. Six papers with five citations, still five. Six papers with six citations, you go up to an H-Index of six.

Those looking to inflate their H-Index can easily do so if they are able to publish regularly – self-citations are often excluded in more comprehensive analyses, but not in most of the automatically generated values. However, that kind of engineered 'citation inflation' requires a regular stream of published papers, and that in turn requires a regular stream of insight generated via new work. Access to a colleague with research funding can facilitate this, but usually only for short-term contracts of two or so years. It's exceedingly difficult to plan a life around institutional and structural career instability. The emotional toll of this system is often not discussed, but includes systemic depression, mental health issues, and increased rates of illness due to stress. Researchers report difficulties in balancing work and family life, the gradual erosion of vacation time, and long, unsustainable working hours.

Talking about these issues is difficult for many – the system as it currently stands is not geared up to seriously consider the emotional toll of short-termism in research planning. A full solution to the problem would be expensive, and require root-and-branch reform of academic promotion structures; redesign of the supply and demand of postgraduate researchers; and reevaluation of the way in which research funding is competitively allocated. It is easier in many cases to simply ignore the problem and hope that the professional consequences of disclosure temper revelatory desires. That is not to say the problem is entirely unacknowledged – welcome steps are now being taken to openly discuss these issues. For those already suffering the emotional toll, it is often too little and too late.

Discussing these issues from a position of career instability is risky and requires a considerable degree of personal bravery – not only to admit that you need help but also to 'speak truth to power'. Power differentials are a common feature of the workplace environment, and those in positions of authority may, or may not, be aware of the suppression effect those differentials may have on those around them. It is clear from our case study that Blackbriar is not reticent in wielding the differential in his favour. For many employers, they may simply be so distant from the day-to-day impact of the issue that they are unaware that it is choking off dissenting voices. Career stability and professional security can be moderators of this problem, but neither of those traits can accurately be ascribed to those on a succession of research contracts. This is especially true when every extension to the contract is dependent on the will and desire for colleagues to seek additional funding for later projects. As with many things, some progress has been made in this area to improve, at least legally, the tenuous position of those who are on successive fixed-term contracts. However, institutions have been as quick to respond through the use of punctuated contracts or zero-hour contracts to ensure that their legal obligation does not stifle their organisational flexibility. As a consequence, postgrads and postdocs can come to seem like interchangeable resources.

The desire to distinguish oneself can lead both to working long beyond what could be considered reasonable hours but also for the willingness to work those unreasonable hours to be seen as a pre-requisite deliverable of scholarly dedication. After all, if I can choose between two otherwise equal research staff members where the difference is their willingness to work weekends, why wouldn't I pick the one that 'goes the extra mile'? With that in mind, how willing might a transient researcher be to rock the boat by kicking up a fuss about the quality of the analysis being performed on data sets?

Principal investigators too are under enormous institutional pressures, and one shouldn't underplay how valuable those 'free' hours can be to a project. Similarly, when it comes to providing full recognition of contribution, the dynamic can lead to a kind of Matthew effect (Merton, 1968). Those with the most power and professional reputation tend to accumulate even more power and reputation because of their ability to shut out or override the concerns of transient researchers. It is not uncommon, for example, for the principal investigators on research projects to insist that they are afforded first author status on all papers generated as a result of their project. This may be required even if their actual contribution was minimal. Most journals and academic outlets have strict rules on how authorship is to be decided but such policies and procedures must always work on the honour system.

This set of interrelated issues make seeking employment in academia a high-stress game of obtaining research funding, rolled into the dominant paradigm of 'publish or perish', within an environment where demand is vastly oversupplied and long-term career stability may be as limited as 'the next three years'. Even that is a stability that may be luxurious for some – the university of the authors often employs adjunct faculty on a '1+1' basis, which is to say a single year with the **possibility** of a year's extension.

Internationally all of this has created an environment where academia is often no longer considered a viable and attractive career destination, with a resultant brain-drain to the private sector. Some in America for example argue persuasively that the position of university professor is no longer considered to be a middle-class profession. The American academic system in particular is full of examples of adjunct professors living on food stamps and picking up a few classes per semester to scratch out a subsistence wage. While it is true that many of these adjuncts are working in fields where employment prospects across the board are weaker, it doesn't change the fact that those considered expert enough to teach and research are not always considered worthy enough to offer legitimate career stability. The effect is to create an ongoing, inexorable attrition where your willingness to play your hand against a stacked deck is as important as your ability to do the job itself. There is little economic incentive for a principal investigator, or even an institution, to address this head-on. The cost associated with exacerbating a condition of long-term burn-out won't be felt within the limited constraints of a single research contract. By the time the worst of the mental toll will be felt, it's highly likely it'll be someone else's problem entirely.

SOCIOLOGICAL AND PHYSICAL ACCESSIBILITY

Another issue raised in this article is that of accessibility – the degree to which technology, facilities, and information can be used by people with extraordinary requirements. Generally, this can be further broken down into two categories of accessibility – sociological accessibility, and physical accessibility. These terms apply equally to all kinds of modern resources, from government services to computer programmes to vehicles. Within this section, we will primarily refer to the accessibility of computer programmes and research data sets, as it is that aspect of the concept that is more relevant to our purposes.

Whether due to cultural constraints, perceived stigma, or general disinterest, it is often the case that certain technological and societal trends are not accorded equal value within different groupings. The degree to which factors internal to these technologies and trends permit generally equal participation defines its sociological accessibility. The way in which certain things are presented or contextualised however can greatly impact on how an individual chooses to perceive their worth. Technology is rarely truly apolitical and often demonstrates the underlying cultural assumptions of its creators (Czaja & Lee, 2002; Keates & Clarkson, 2003). Consider for example the issues of male versus female wish fulfilment represented by many video games (Heron et al., 2014), and consider how the way in which explicitly gendered protagonists may appeal, or otherwise, to groups of men and women. Consider the cultural connotations of colour choice in children's toys. Traditionally, this is blue for boys and pink for girls. Consider how that impacts on the choice of early play for both children and their parents, and the stigma that may be experienced by obviating cultural norms.

More pertinent to this particular case study is the issue of physical accessibility. When someone has overcome whatever sociological barriers may have been in the way, and actively wishes to engage with technology, then the extent to which that technology is usable is pertinent. While things have gotten much better in recent years, it is still the case that software is often inaccessible to people with physical and mental disabilities. Blind users often find screen-reader technology works well for the most part. However, such technology has occasional missteps as a result of software not being designed to work with standard tools such as JAWS, and with internet conventions such as subverting alt-text or communicating via emojis. Users who are colour blind may find that they are unable to distinguish visual cues when the only differentiating factor is the colour (for example, a green cursor that turns red, or red warning text). Where sound is used to deliver important interface information, deaf users are often disadvantaged, especially when viewing videos without subtitles. Users with mobility impairments may find that software requires too much fine-grained movement, or too many simultaneous key-presses, or is simply tiring to use with non-standard interaction devices such as head-wands or mouth-sticks. The more intensely interactive a piece of software is, the more these issues become important and the more difficult it becomes to truly compensate for all interaction regimes. For most desktop software packages, interaction is not intense and does not come in short bursts. For these packages, inaccessibility tends to be an oversight or as a consequence of a lack of awareness of the issues. However, for some highly interactive software packages, such as video games, the problem may be more difficult to fully address (Heron, 2012). It is a problem not confined to software though – physical products too manifest many inaccessibilities (Heron et al., 2018a, 2018b).

Large corporations can afford to have dedicated developers whose sole job is to work on software accessibility. Small projects, open-source or otherwise, tend to be mostly auteured-driven products. Research software in particular, especially that written for a particular project or research team, can rarely muster development time beyond the lifespan of the funding (Heron et al., 2013a). Some systems are primarily written to test a concept or fulfil an immediate need. Such software is only incidentally accessible unless explicitly written for the purposes of accessibility research (c.f., Heron et al., 2013b, 2013c).

In our case study, our postgraduate student has encountered an inaccessible piece of research software. The additional logistical overhead this creates has put restrictions on his operating flexibility. He could only access the research data that other people could provide for him. We already know from earlier articles in the study that the set of people who had any access to the data was highly restricted. Such constraints make full, effective oversight of data and its analysis extremely costly, extremely time-consuming, and likely to be de-emphasised in an environment where more pressing needs had to be immediately serviced. We know that the postgraduate students were engaged in teaching, seminar work, and marking for Blackbriar – all of these come with deadlines, oversight, and committee work. We are also told that they were expected to be 'always available' to service his research needs. Supporting a colleague doesn't seem like it would have been a high priority given what we have been told. Ensuring the utmost veracity of already trusted data may have been a luxury that simply could not be afforded in the context of the working environment.

There is much that can be done to ensure accessible software but it requires both the will to invest the effort and the skillsets to make structural changes to the underlying programming code. In many research environments, we cannot assume either – software in these circumstances is not being written as a production-ready product. It is instead intended to serve as a stop-gap solutions that meet an immediate, but likely nonpersistent need. Sometimes, temporary software solutions evolve into a core element of an organisational workflow, but there are inertial pressures that come into play when the need comes for changes. Sometimes software is so tightly bound up in its original assumptions that making an adjustment requires a complete rewrite of the code. Sometimes the source-code was only ever stored in the personal directory of a postdoctoral researcher who left the institution ten years ago. Sometimes the institution is no longer subscribed to the development tools that were used to create a software solution. Sometimes the operating context moves on, and updating to the newest software libraries breaks compatibility.

When making improvements for accessibility purposes, there is a relatively specialised skillset required to make sure that changes don't have an overall negative effect (Milne et al., 2005). There is always a reason why something shouldn't be done, and with the maintenance of software the reason may be 'nobody can actually do it'. The fewer people impacted by a problem with the software, the less likely it is that the need for change will gather sufficient urgency or develop the critical mass to turn 'this should be done' into 'this is being done'.

With larger software suites which are purchased from commercial outlets, we may also be restricted to what can be done within the context of an established extra-institutional user-base. There may be thousands, or tens of thousands, of users who all have their own views on how the software should be improved. A larger user-pool would mean that accessibility issues were experienced by a larger number of people, but they are likely still only a subset of the installed user-base. Economic self-preservation will always come into play. A company looking to profit from its users must see to the needs of the many before it can justify seeing to the needs of the few. It's possible to marshal any number of moral and ethical arguments as to why software should be made accessible as a first priority, but such arguments may not convince an organisation dealing with the ongoing triage implied by competition within complex and unpredictable economic restrictions. In balance of this, most countries have laws that mandate core materials and systems should be accessible, but enforcing compliance is difficult.

Thus, we see situations like this where people 'find a way' around the issue, often by following a tortuous chain of importing and exporting until the right data can be presented in the right way with the right level of accessibility. Such compensations are invariably fragile – if any part of the compensatory chain is altered, the entire thing may fall apart. Software changes on a regular basis if it has active developers, and these changes can often be substantial. They may include things such as changing entirely the default interaction metaphor; dropping support for whole families of tools; or removing the ability to import or export particular formats. Sometimes the impact is almost dystopian, such as when Adobe required users to pay an additional subscription to use a certain family of colours. Each time a compensatory process is broken, it takes time to repair. It is rare there is no route to accessibility through these kinds of improvised solution spaces, but that too occasionally occurs. Here, we must ask ourselves to what extent Duncan can be held responsible for not accessing data hidden behind layers of accessibility obstacles.

THE UNEQUAL DISTRIBUTION OF DOMESTIC LABOUR

Sharon McAlpine raises an important point here in this article. It's an aside, but it carries a lot of weight. Everyone was impacted by the pandemic. Global disruption was on a scale unprecedented in modern memory. That disruption though didn't fall equally across all demographic categories. Some suffered more than others. And in that, sometimes all that is needed for one group to get ahead is for another to fall behind.

Research conducted around workplace patterns during Covid-19 have revealed several alarming trends. Demographic sectors reporting the largest challenges to a shift in working patterns included working mothers; women in senior management; and Black women. When women had younger children – under ten in most cases – the attrition rate associated with 'downshifting' their career was ten percentage points higher than for equivalent men (Bluedorn et al., 2021). More women than men lost their jobs (Brodeur et al., 2021) and more women than men were in the kind of 'essential jobs' that were public facing and thus most at risk

(Brugiavini et al., 2022). Women experienced greater rates of work disruption due to increases in the need for childcare and other domestic responsibilities (Wenham et al., 2020). It wasn't all doom and gloom. The pandemic did also result in men contributing more to the largely unpaid and unrecognised 'care economy' (DeRock, 2021) that is overwhelmingly enabled through the efforts of women. When such a shift happens though, additional household responsibility tends to accrue disproportionately in the activities considered to be more enjoyable and with greater social approval such as childcare (Dinh et al., 2017). Less visible, less enjoyable domestic work rarely records equivalent change. It has long been argued by scholars of gender issues that there is a need for Gross Domestic Product (GDP) to reflect the value of care work and domestic labour – the lack of easy (even if misleading) quantification hides much of the disproportionate burden taken on by women during the pandemic (Power, 2020). A full accounting of this topic is outside the scope of this chapter, but we do recommend you check out the masterful and scholarly book *Invisible Women* (Perez, 2019) for a much fuller – and much more convincing – accounting.

For McAlpine, working within a university system, she may have been interested in the social media trend that rose up during the early days of the pandemic – recognition of the growing #CoronaPublicationGap. As the responsibilities upon women increased, the productivity of male colleagues rose in comparison. This only served to exacerbate existing disparities – Oleschuk (2020) provides a compelling list of these, including presence of women in faculty positions; publication rates; citations; recognition; and salary. The same paper notes that these are not reflective of differences in capability, but rather a result of systemic injustices. In other words, the Coronavirus created a situation where the rich got richer, at the expense of the poorer. Women professionals in other fields experienced a wide range of negative pressures – mental health was seriously impacted, physical pain increased, eye-strain was exacerbated, and the combined effect of this was an increase in emotional deregulation (Sharma & Vaish, 2020). Others have self-reported significant negative impact on social activities, with increased loneliness as a reliable result (Giles & Oncescu, 2021).

Sharon's comment then is not simply a throwaway complaint but reflective of a trend that was experienced widely throughout the shift to remote working. Academic systems already give great advantage to those that can devote larger amounts of time to their vocation, and men as a whole have more time to spend on work and leisure. That time too tends to be less fractured by other calls than that of equivalent women (Dinh et al., 2017; Perez, 2019). It is difficult to put hard numbers on activities that are, perhaps purposefully, excluded from economic evaluation. If you ask men and women about their perception of productivity and happiness during the pandemic though, they show significant differences (Feng & Savani, 2020). GDP itself is a gendered construct, and one that explicitly prioritises the 'traditional work' of men (Perez, 2019) and views the unpaid work of the care economy as a 'costless resource' (Himmelweit, 2002). As some commentators have noted, one has to wonder if the skewing of these figures is intentional. If we knew how much it cost, someone would be expected to pay for it.

McAlpine also adds something else of importance here – she identifies the luxury of space that is assumed of those working from home. As previously mentioned, some of us have flats large enough to support home offices and all the comforts needed to engage productively with our careers at a distance. Others must find an unused corner of a bedroom or share a kitchen table with family members. A remote lecture is easy to give from a dedicated room with the requisite software. It can be borderline impossible if trying to do so when family members keep wandering by in the background of a living room. Sharon notes that since her husband had the 'high-powered career' he was the one who got the pick of working location. Single-purpose rooms became dual-purpose work-spaces during the pandemic. The quality of the experience of working at home depends intensely on the exact configuration which people can arrange to do it (Watson et al., 2021). Some forms of work require deep concentration over long periods of time – some can be dealt with more incrementally. Some households suffered with the shift to relying on software and hardware they may not have previously used. Before Covid-19 hit, there were significant numbers of people who had never heard of Zoom, and had never owned a webcam. Getting the equipment, due to previously mentioned supply-chain issues, could be difficult. Installing it in a robust way could require specialist skills. Ergonomics of the working environment can be hard to control without a significant budget or without financial support from an employer. Even when dedicated spaces were available, the presence of young children kept home from school could create difficulties in maintaining distance between home and work (Wethal et al., 2022). Young children for example may not understand why mummy won't come out of her room to play with them. All of these things are hard enough to navigate **without** a partner who believes that professional work and additional home work can be combined without undue stress.

Blackbriar also seems not to have helped, expecting his own workaholism (Golden, 2012) to be reflected in the relationship his students have with their own professional practice. The perception of overwork in the academy (Jacobs & Winslow, 2004) is one that is often reflected in the joke, 'Academia is very flexible – work whatever eighty hours you like'. Much of this is due to norms that are difficult to shift, but also due to a shared delusion as to how much benefit comes from additional hours of work despite the evidence being quite the opposite (Pencavel, 2015). Changed life circumstances, such as pregnancy, can break the chain between the illusion and the reality (Lupu, 2021). Larger, more systemic change seems harder to achieve. Again, the impact of excessive expectations on working hours differs by gender – women who don't want to work 80-hour workweeks tend to seek accommodation at the cost of their professional reputation. Men will seek ways to covertly turn an excessive work week into a more manageable one to retain the benefit of being thought of as reliable 'superstars' (Reid, 2015).

How much do you need to work to have a successful academic career? Opinions, like in every field, vary. Suffice to say though that the evidence suggests that those working 80-hour weeks are less common than they claim, and those that do are producing less with those additional hours than they might like.

THE NATURE OF FEEDBACK

What then of the claim that Blackbriar had largely given up responsibility for writing his supervisory notes and had turned to generative AI? Or that Sharon McAlpine had used an AI chatbot to help with the writing of a paper? At the time of writing, these are exceptionally hot-button issues in computer ethics. It will be some time before the field approaches anything that could be described as a consensus. However, just because it's complicated doesn't mean we should avoid it – we're always going to be working in uncharted waters in this area. The field moves too fast for us to always have comfortable answers.

Let's pull back a little from the specifics here and focus on the intention – is anything lost in either of these cases when it comes to the use of AI-supported writing? The answer is… maybe! We don't know enough to say, but we do know what people would be hoping to get out of the text that came from the machines.

We should perhaps though draw a distinction between the different forms that writing takes. Occasionally, Michael will be 'caught' playing a video game when he's supposed to be writing. Upon being asked 'I thought you were writing?', he'll often answer, 'I am'. The mechanical act of 'writing down' is distinct from the more thoughtful activity of contemplating what should be written, and that sometimes requires refocusing on a routine task so as to relieve the mind of other distractions. For Michael, the act of engaging with a repetitive game-play loop is part of his process. Or so he claims.

Turning over the responsibility to 'Discuss the conclusions of this paper' to an AI is unlikely to yield – at the time of writing – thoughtful results. However, asking an AI to turn a raw thought into a more polished form can be an efficient – and effective way to blend the benefits of human and machine output. We don't question whether the use of a thesaurus changes the authenticity of writing. We do not object to spellcheckers. Even tools that provide writing prompts and can complete sentences have become commonplace. At what point does this computer-supported creativity become an unacceptable violation of our trust?

We've already talked about some of the issues that come along with the inclusion of generative AI – particularly in relation to how the content is probabilistic, and that it doesn't necessary reflect the facts underpinning the real world. The output of GPT can sometimes resonate eerily, as if it comes from an alternate universe a few microns away from our own. That said, with human input at the start and human editing at the end, is there really a problem if the act of composition is offloaded onto machines?

Perhaps yes, perhaps no. Realistically it seems that we are likely to see more and more of this happening in the next few years, but what AI routines regularly fail to do is manage to capture anything of the human soul. The writing, while competent, is often lacking in anything approaching a sense of style. The content lacks deeper insight. While it can integrate diverse perspectives, it also can't really take a side in a way that is compatible with evidence. AI writing is not thoughtful, because there is no thought available. Prompt engineering – the process of getting usable results out of generative AI – is likely to be a growth area over the short term for this reason. That is, of course, until AIs that generate good prompts render that specific need obsolete.

Perhaps then the question is, 'how much thought went into prompting the AI?' as opposed to 'How much of the text was computer generated?'. We have some hint of that in the comment that Duncan made regarding the performance review that would lead to attendance at a research retreat – a claim that no one had made the grade later became a revelation that the professor hadn't even looked at the work submitted. This has shades of the claims made regarding the management culture at ZA/UM, responsible for making the game Disco Elysium game. Presumably what McAlpine and Duncan wanted was thoughtful feedback from their supervisor. The un-nuanced output of an LLM seems unlikely to provide that in any meaningful sense.

Sharon's use of an AI tool to generate some discussion text mostly falls within the same frame except that it's a little more complicated because of the formality associated. Journals, book publishers, conference chairs, schools... all of these have come increasingly down on the use of generative text and images within publications and assignments. Whether Sharon is at fault here largely depends on the terms and conditions under which she submitted the piece, and on what transparency was provided to the journal editors. Some GPT text is used within **this** book for example, but it is always clearly identified and provided as part of a commentary on generative AI. There is no intent **or attempt** to deceive. It is illustrative only. If there's no context such as 'We used GPT to generate this paragraph', then the text was written by us. As such, it is within scope and appropriate, and does not violate the contract under which this book was produced. Is the paper submitted to the journal similarly inoculated against allegations of misconduct? We don't know – and indeed, we'll never know – but it's possible a revelation like this could have larger repercussions on the authors than might be expected. Is it grounds for a retraction, or for a corrigendum, or for nothing at all? Only time can tell.

REFERENCES

Bluedorn, M. J. C., Caselli, F. G., Hansen, M. N.-J. H., Shibata, M. I., & Tavares, M. M. M. (2021). *Gender and employment in the COVID -19 recession: Evidence on "she-cessions"*. International Monetary Fund.

Brodeur, A., Gray, D., Islam, A., & Bhuiyan, S. (2021). A literature review of the economics of COVID-19. *Journal of Economic Surveys*, *35*(4), 1007–1044.

Brugiavini, A., Buia, R. E., & Simonetti, I. (2022). Occupation and working outcomes during the coronavirus pandemic. *European Journal of Ageing*, *19*, 863–882.

Czaja, S. J., & Lee, C. C. (2002). Designing computer systems for older adults. In *The human-computer interaction handbook: Fundamentals, evolving technologies and emerging applications* (pp. 413–427). CRC Press, Inc.

DeRock, D. (2021). Hidden in plain sight: Unpaid household services and the politics of GDP measurement. *New Political Economy*, *26*(1), 20–35.

Dinh, H., Strazdins, L., & Welsh, J. (2017). Hour-glass ceilings: Work-hour thresholds, gendered health inequities. *Social Science & Medicine, 176*, 42–51.

Feng, Z., & Savani, K. (2020). Covid-19 created a gender gap in perceived work productivity and job satisfaction: Implications for dual-career parents working from home. *Gender in Management: An International Journal, 35*(7/8), 719–736.

Giles, A. R., & Oncescu, J. (2021). Single women's leisure during the coronavirus pandemic. *Leisure Sciences, 43*(1–2), 204–210.

Golden, L. (2012). *The effects of working time on productivity and firm performance, research synthesis paper.* International Labor Organization (ILO) Conditions of Work and Employment Series No. 33. Conditions of Work and Employment Branch.

Heron, M. (2012). Inaccessible through oversight: The need for inclusive game design. *The Computer Games Journal, 1*(1), 29–38.

Heron, M. J., Belford, P., & Goker, A. (2014). Sexism in the circuitry: Female participation in male-dominated popular computer culture. *ACM SIGCAS Computers and Society, 44*(4), 18–29.

Heron, M. J., Belford, P. H., Reid, H., & Crabb, M. (2018a). Eighteen months of meeple like us: An exploration into the state of board game accessibility. *The Computer Games Journal, 7*(2), 75–95.

Heron, M. J., Belford, P. H., Reid, H., & Crabb, M. (2018b). Meeple centred design: A heuristic toolkit for evaluating the accessibility of tabletop games. *The Computer Games Journal, 7*(2), 97–114.

Heron, M., Hanson, V. L., & Ricketts, I. (2013a). Open source and accessibility: Advantages and limitations. *Journal of Interaction Science, 1*(1), 1–10.

Heron, M., Hanson, V. L., & Ricketts, I. W. (2013b). Access: A technical framework for adaptive accessibility support. In *Proceedings of the 5th ACM SIGCHI Symposium on Engineering Interactive Computing Systems (EICS '13)* (pp. 33–42). Association for Computing Machinery, New York, NY. https://doi.org/10.1145/2494603.2480316.

Heron, M., Hanson, V. L., & Ricketts, I. W. (2013c). Accessibility support for older adults with the access framework. *International Journal of Human-Computer Interaction, 29*(11), 702–716.

Himmelweit, S. (2002). Making visible the hidden economy: The case for gender-impact analysis of economic policy. *Feminist Economics, 8*(1), 49–70.

Jacobs, J. A., & Winslow, S. E. (2004). Overworked faculty: Job stresses and family demands. *The ANNALS of the American Academy of Political and Social Science, 596*(1), 104–129.

Keates, S., & Clarkson, J. (2003). Countering design exclusion. In *Inclusive design* (pp. 438–453). Springer.

Lupu, I. (2021). An autoethnography of pregnancy and birth during Covid times: Transcending the illusio of overwork in academia? *Gender, Work & Organization, 28*(5), 1898–1911.

Martin, B. R. (2011). The research excellence framework and the 'impact agenda': Are we creating a Frankenstein monster? *Research Evaluation, 20*(3), 247–254.

Merton, R. K. (1968). The Matthew effect in science: The reward and communication systems of science are considered. *Science, 159*(3810), 56–63.

Milne, S., Dickinson, A., Carmichael, A., Sloan, D., Eisma, R., & Gregor, P. (2005). Are guidelines enough? An introduction to designing web sites accessible to older people. *IBM Systems Journal, 44*(3), 557–571.

Oleschuk, M. (2020). Gender equity considerations for tenure and promotion during covid-19. *Canadian Review of Sociology, 57*(3), 502.

Pencavel, J. (2015). The productivity of working hours. *The Economic Journal, 125*(589), 2052–2076.

Perez, C. C. (2019). *Invisible women: Data bias in a world designed for men.* Abrams.

Power, K. (2020). The COVID-19 pandemic has increased the care burden of women and families. *Sustainability: Science, Practice and Policy, 16*(1), 67–73.

Reid, E. (2015). Why some men pretend to work 80-hour weeks. *Harvard Business Review*, 28.

Sharma, N., & Vaish, H. (2020). Impact ofCOVID-19 on mental health and physical load on women professionals: An online cross-sectional survey. *Health Care for Women International, 41*(11–12), 1255–1272.

Watson, A., Lupton, D., & Michael, M. (2021). The COVID digital home assemblage: Transforming the home into a work space during the crisis. *Convergence, 27*(5), 1207–1221.

Wenham, C., Smith, J., & Morgan, R. (2020). COVID-19: The gendered impacts of the outbreak. *The Lancet, 395*(10227), 846–848.

Wethal, U., Ellsworth-Krebs, K., Hansen, A., Changede, S., & Spaargaren, G. (2022). Reworking boundaries in the home-as-office: Boundary traffic during COVID-19 lockdown and the future of working from home. *Sustainability: Science, Practice and Policy, 18*(1), 325–343.

CHAPTER 7

Leaked Minute Lays Bare University Culture

NEWSPAPER ARTICLE

An exclusive report by Jack McKracken for the Dunglen Chronicle

A confidential minute has been leaked to the Dunglen Chronicle, revealing the culture of research that was being enforced by the Vice-Chancellor, Professor Sir David Tumblewood. The minute is from a closed meeting with the 'Professoriate' – the group of professors within the university, and is from a meeting that he chaired shortly after being instated as principal of the university ten years ago. The minute documents the principal's feelings on research, the importance of offering 'value for money' for funding partners, and how important it is that the university finds ways to propel itself up the league table and return better results to the Research Output Evaluation Framework Committee (ROEFC). League tables have become increasingly important in higher education as an easy to understand way for those outside academia to identify which is the 'best' institution. The ROEFC is a government-mandated assessment exercise that evaluates the value and impact of research done within each university. League tables are a hugely important recruitment tool for universities. A good result in the ROEFC can ensure access to larger research grants.

DOI: 10.1201/9781003426172-7

Ten years ago, the University of Dunglen was languishing at the tail end of the league tables – as a new university, one established in 1992 as part of the expansion of the sector; it struggled to achieve much progress in either metric. Placement in the league tables is partially dependent on the quality of research output, and when Professor Sir David Tumblewood was appointed, he made it clear to the Professoriate what he expected. Quoting from the minute:

```
The principal reminds all present that research is one of the most effective levers to build academic credibility. Academic credibility in turn builds research capability as new funding streams and partnerships become available. There is a positive feedback loop here we must exploit. As the university becomes known for the quality of its research, this will increase our standing in the league tables which will ensure the best students come to us.
```

Since we currently lack the ability to gain significant amounts of new research funding, we must ensure that the funding that we do have is fully utilised. Research resources such as lab equipment, research students, and postdoctoral fellows are valuable and currently underused. We must seek to increase the amount of useful value we get out of these resources, as that is the easiest way for us to increase the quality and quantity of our research outputs in the short term.

The principal also reminded everyone present that their individual research appointments were provided on the basis of sustainability, and that sustainability required the procurement of external funds. Those that could not source such funds would be at risk of their professorial appointments being downsized.

The principal put the expected income of a research professor at 'enough to fund all postdocs and postgraduates at full economic costing, plus 30% overhead for the university'. The Chronicle contacted Emeritus Professor Joanne Clement, a recently retired employee of the university. She spoke on the record about the culture of research that the new principal had created.

'I was a professor in the anthropology of volunteering. I spent my time studying the reasons why people volunteered, how they did it, and what they got out of it. This had previously been a topic the university valued quite highly as part of our civic engagement duties. The previous principal, Professor Elizabeth Burke, believed that we had a responsibility to the community embedded in our university charter. The more financially successful departments in the university sort of subsidised mine – that's what the overhead is supposed to do, after all. Some research is important, but not "sexy" enough that it'll get enough funding. My area was that.

I was told point blank by Tumblewood that if I couldn't secure enough research funding for my small team of four postgrads and two postdocs, that my position would be eliminated and I'd either have to accept a demotion to senior lecturer or retire. The demotion would have basically removed any of my ability to do research as I would go from teaching six hours a week to teaching almost sixteen. Since I was only a few years off retirement anyway, I chose the latter. They let me keep my professor title, since it didn't actually cost them anything. That is why I am an "Emeritus" professor, but even that's at the gift of the university and subject to the same politics as anything else. There were those in the university, Blackbriar amongst them, who said that my research had never really been of the necessary standard for the title, and that allowing me to keep it would reduce the perceived value of their own titles'.

The Chronicle understands that 12 research groups within the university have been shut over the past ten years as a result of professors being unable to secure reliable streams of funding. These include the Centre for Childhood Studies, the Institute of Professional Ethics, the Centre for Equality, the Truth in Journalism Institute, and the Refugee Study Group. 'This is all part of a larger push towards embedding "business thinking" in academia', Clement says. 'Sometimes a university has to do the things that aren't financially viable because nobody else will do them. Work can be important without being valued by those who have the money to fund it. The Truth in Journalism Institute is an excellent example of that – we'd all like more truth in journalism, no offence, but the only people with the money to pay for it have a vested interest in the press being easy to control'.

Clement, in follow-up emails, talked about what she saw as the decay of modern academia. 'Once upon a time our job was to inspire wisdom. Now it's to produce information. We've lost sight of what universities are supposed to be – places where we actually think about things. Most of the research that universities produce is about bibliometrics – publish often, in as high an impact venue as possible, to make the numbers go up. In the process, the entire sector has become a "write only" industry. People rarely reflect upon the wider context of what's happening, because the only way to survive in academia is to package up information into consumable McNuggets. University professors used to have a big idea and spend their careers pursuing its implications. Now they load up a shotgun with gimmicky papers and fire it at a conference in the hope some get through. We've stopped being thoughtful, and that's reflected in what happened at Dunglen'.

It wasn't just research centres that closed, Clement explains. Any professor who had no formal research centre but wasn't bringing in the grant money to support their own position was made redundant. Similarly with those who had the position of Reader, an academic title broadly half-way between senior lecturer and professor – if their funding couldn't be externally supported, their position was eliminated.

However, Professor Derek Taylor who currently heads the Dunglen Business School offers a counter argument. 'Like it or not, the days of the university ivory tower are gone. If you aren't producing value sufficiently obvious for someone, somewhere to pay for it then it can no longer be subsidised. Universities are more accountable to the public now than they ever have been. Besides, look at the results that have been achieved over the past ten years – what further validation of Sir David's philosophy do you need? Universities are a business. Research is a business. Anyone who believes otherwise these days is living in the past'.

The University of Dunglen has indeed enjoyed something of a golden age in the past decade. From an average position of 76 out of 117 universities ten years ago, the last league table positions put Dunglen at 32. In 2014, Dunglen had a ROEFC profile of 1.3, suggesting only national importance in terms of research outputs. For the last exercise, this had increased to 3.1, suggesting the majority of output was considered 'internationally excellent' with the occasional instance of 'world leading'. The consequence of this is that a much larger degree of research funding is available to the university, and success rates in competitive bids have increased exponentially. This comes both directly from governmental grants and from commercial partners attracted to a growing reputation for research quality. Researchers from all over the world look to come to the University of Dunglen to develop or enhance their research profiles.

'Ten years ago our research was very weak', says Taylor. 'Now it's regarded amongst the best in the world. We've not only increased the quality of the research we do, but we've also increased the quantity by an order of magnitude. It may have been a painful process to get to here from there, but the university is much stronger, much healthier and much more respected internationally as a result. Our students benefit from this immensely – student employability is directly linked to institution prestige. Everyone wins'.

Clement again has a contrary opinion. 'We value what we measure', she said in an email. 'We do not measure what we value. Any time a quality metric becomes a target, it ceases to capture quality. Academia is a game, and while I don't see a way that younger colleagues can stop playing I think the rest should be doing what they can to help the sector rediscover its public-facing mission'.

FUNDING AND PROFESSIONAL SECURITY IN HIGHER EDUCATION

The leaked minute has stirred things up further, but we need to spend a little time contextualising what we're actually being told. Understanding context is the first step to appreciating nuance. Funding structures, particularly within the United Kingdom, can be complex to unpick. Although we often talk of the UK Higher Education (HE) sector, this is somewhat misleading. The UK is made up of four component countries: Scotland, England, Wales, and Northern Ireland. Scotland, as a consequence of the Act of Union (Arnott & Menter, 2007; Mooney & Poole, 2004), retains control over certain elements of its national public provision – Education, Law, and Religion are chief amongst these. In addition to this, via the devolved Parliament in Holyrood, Edinburgh exerts control over numerous other aspects of Scottish governance. The money to fund this work is derived from what is known as the Block Grant – an allocation of funds to the devolved parliament for it to budget according to local priorities. This funding comes from Westminster but is allocated according to Scottish revenues and tax receipts as a proportion of public spend in England. Within England and Wales, as a consequence of the 2014 Browne Report (Browne, 2010), university teaching is funded primarily through student fees. Certain courses deemed to be within the public interest, such as Medicine, are also subsidised by Westminster.

Within Scotland, Scottish students do not pay fees for their higher education as the provision of such is, at least at the time of writing, deemed to be a public good for the benefit of society as a whole. Funding for universities in Scotland, where the fictional University of Dunglen is situated, is allocated by Holyrood according to national and political priorities.

This creates a complex funding context when reporting on higher education in the United Kingdom. Popular discussion is often flavoured by a lack of appreciation of the significant differences in student expectations and provision. There is a considerable difference in assumption between students in a system driven by marketisation (England, America) versus one where access to HE is still considered to be a public good (Scotland, Sweden). We would have much to discuss that is different were Dunglen situated in England simply as a consequence of student attitudes. As alluded to above, none of this nuance is present in the popular accounts of the scandal as reported in the newspaper. This may be a matter of necessary concision; or a consequence of the assumption of knowledge upon the reader; or it may be a deliberate attempt to influence the presentation of the story. We simply cannot know and as such we must again come back to our grounding principle – be wary of placing too much emphasis on a single point of narration.

The situation is complicated further by the perceived bifurcation of university interests into 'teaching' and 'research'. Funding for the latter is even more complicated, coming as it does from multiple sources. One category of sources is the 'Higher Education Funding Councils', including HEFCE in England and the Scottish Funding Council (SFC) in Scotland. These allocations are provided as blocks to individual institutions, but these are not awarded equitably – instead, the amount particular institutions receive is directly influenced by their performance in the Research Excellence Framework (Martin, 2011). This is an assessment exercise into the impact of the research in UK universities and is conducted roughly every five years under one name or another. Those universities that have done well in research within their funding period are likely to receive the lion's share of this block funding, with the others receiving comparatively little. Of an approximately £294.30m Scottish dispensation in 2022,[1] Robert Gordon University received an indicative settlement of £2.1m. Aberdeen, in the same city, received £23.9m. Dundee University received £21.3m. Abertay University, also in Dundee, received £1.3m. Edinburgh university received £91.1m. The obvious impact of this dispersal of funds is to create a kind of funding Matthew Effect (Merton, 1968), ensuring again that as a general rule 'the rich get richer'.

We're seeing that principle a lot play out in this case study.

With this in mind, many universities look to supplement these often miserly research allocations by seeking competitive funding from various research councils. It is in relation to this area that we see the principal of Dunglen threatening his professors. Research-active academics tend to engage in continual grant writing to secure these funds, putting together proposals for programmes of research. They then use the funds that are hopefully generated to do the work that will put the academic, and his or her institution, in a more favourable light for future research fund generation. As in many things, success breeds success. Within the UK, funding may be at a national level through one of the bodies such as the Engineering and Physical Sciences Research Council (EPSRC) or the Arts and Humanities Research Council (AHRC). A certain degree of industry funding from larger organisations with dedicated research departments is also available. Some minor funding too can come as a result of individual ad hoc collaborations. Professor Blackbriar's funding would fall largely into this latter bracket.

The funding situation has become even more complex in the UK with regard to Britain's exit from the European Union. Technically speaking UK researchers can still apply for European funding. Horizon Europe is a huge initiative, responsible for dispersing around 95 billion euros of funding, and runs until 2027. The UK holds an associate third country status within this programme, paying access fees to secure funding opportunities on an equal basis to EU members. There is anecdotal evidence to suggest that even while the UK is a third country member with access to this funding, there is always a political element when UK universities are part of the consortiums seeking money.

Considerable sums may potentially be available from these funding bodies. In 2022, Horizon Europe gave out a total of 1.3bn euros over 665 projects. Hidden in that success story is the failure cost – only 12% of submitted applications received funding. Someone applying for a starting grant from the European Research Council in 2022 could expect to see only a 16% chance of success. This lack of return on invested effort is perhaps responsible for the fall in applications – the 2,932 applications received in 2022 is a fall of 28% from 2021, and a fall of 10% from 2020. National funding bodies exist to supplement this European funding, allowing for local priorities to be addressed with more agility. The funding rates here are sometimes better, but often worse. One bid submitted by the authors of this book went to a call that had a 6% funding rate. Even for those academics that are successful in obtaining grants, the funds they are awarded are not the funds they have available for research. Universities claw back a significant proportion of these to fund other activities and provide support for shared research spaces and organisational infrastructure.

It is in this context that we must consider the minute from the Principal's Office. Those attaining the rank of professor often do so after a sustained history of successful grant generation – this is sometimes formally acknowledged, and other times tied to a nebulous definition of 'excellent research performance'. The London School of Economics includes research grants as evidence of the latter. Kings College London has explicit expectations of funding generation including for lecturers. Lecturers are expected to annually generate £40,000–£50,000 of funding, whereas professors are expected to bring in £150k–£200k. Within the Scandal in Academia, we are told that professors are expected to cover the 'Full Economic Costing' (FEC) of their entire cohort of postdocs and postgraduates, as well as provide on top of that an additional 30% for the university. This is not an uncommon situation, and the pressures that this exerts upon individuals struggling to meet these criteria can have tragic consequences. Imperial College in 2014 placed Stefan Grimm, professor of toxicology, under considerable stress because of his perceived poor grantsmanship. Some of the tenor of this is indicated by the following email that was sent to him by his head of department:[2]

> I am of the opinion that you are struggling to fulfil the metrics of a Professorial post at Imperial College which include maintaining established funding in a programme of research with an attributable share of research spend of £200k p.a and must now start to give serious consideration as to whether you are performing at the expected level of a Professor at Imperial College.
>
> Please be aware that this constitutes the start of informal action in relation to your performance, however should you fail to meet the objective outlined, I will need to consider your performance in accordance with the formal College procedure for managing issues of poor performance (Ordinance D8) which can be found at the following link…

Grimm continues with his own observations:

> On May 30th '13 my boss, Prof Martin Wilkins, came into my office together with his PA and ask me what grants I had. After I enumerated them I was told that this was not enough and that I had to leave the College within one year – "max" as he said. He made it clear that he was acting on behalf of Prof Gavin Screaton, the then head of the Department of Medicine, and told me that I would have a meeting with him soon to be sacked. Without any further comment he left my office. It was only then that I realized that he did not even have the courtesy to close the door of my office when he delivered this message. When I turned around the corner I saw a student who seems to have overheard the conversation looking at me in utter horror.

We see here echoes of the treatment of the fictional Joanne Clement, the Emeritus Professor in the Anthropology of Volunteering. Joanne Clements was merely bitter. Stefan Grimm committed suicide. An inquest in June of 2014 found that he had taken his own life by asphyxiation. We do not, in this book, profess to have any deep understanding of the complex interplay of pressures and personalities involved in this sad situation. Undoubtedly, the truth of the matter is more nuanced than public accounts permit us to appreciate. The basic shape of the incident though is simple enough.

It's a tragic story, and one which is thankfully not often repeated. There are other cases with less dire outcomes, such as professors being dismissed because of a lack of grant generation (Joyner, 2014). Even for those that do not have any direct professional consequence for missing expectations, the effect of the ongoing stress can be corrosive to well-being and mental health. However, the other side of this is a not unreasonable expectation of high performance from those that have been appointed as professors. Universities have a responsibility to get the most out of the money that they spend on their staff, and the idea of meeting targets for income generation is baked into the bones of many professions out in the 'real world'. As with many things in our discussion, there is not a clear path through what is 'right' and what is 'wrong'. Stefan Grimm's case is especially poignant given the outcome, but that can only truly be appreciated with the benefit of 20/20 hindsight. It is safe to say though that the current system, as it stands, is one of potentially extreme personal cost where individual effort is only somewhat correlated with successful bids.

Phenomenal pressures are often put upon research academics by their institutions. The circumstances under which substantial grants may be awarded is often too capricious for any genuinely reliable income to be generated. For those trying to keep a research centre running, the pressure also includes that of securing the jobs of other people, or at least ensuring that there is some degree of progression for those that may be stuck on a succession of fixed-term research contracts. This is not a situation unique to academia – we now live in a world of 'flexible' employment where the conception of a job for life seems almost archaic. A plague of zero-hour contracts is in the process of robbing a significant proportion of an entire generation. Meaningful job security is being pickpocketed wholesale. Universities are a significant offender in this area but they are far from alone in exploiting the situation for their own ends. Those lucky enough to have permanent contracts have security only as long as it is within the interests of the organisations to which they are attached. We pay lip service as a culture to the importance of work-life balance. We live though in a reality where many must work several jobs just to make ends meet and have perilously little scope to exercise employment rights without risk. We focus on the academy within this paper only because that is where the scandal is set. We can see broad parallels to all of this everywhere we look.

THE QUANTIFICATION OF ACADEMIA

Professor Taylor, within the Scandal, notes that universities must demonstrate their value and impact. He notes that it's necessary to embed business thinking into the wider civic role of the academy in modern life. This is an argument often made by critics of universities: that they exist in an ivory tower, disconnected from reality. However, such a hard-nosed economical approach disproportionately threatens much of the good work that is done for its own sake, or for the furthering of human knowledge or culture (Collini, 2012). We need to consider the issue in slightly less emotive and sensationalist terms – the question becomes 'is this work that the public should be subsidising?'.

There are compelling voices on both sides of the debate, and those arguing for prudence in research allocations are also those arguing that it's important to quantify exactly what we get out of research at all. This process has gone by various names over the years – 'citation counting' in the 60s and 70s (Scales, 1976), 'bibliometry' in the 80, 90s, and 00s (Giske, 2008), and 'impact factor' in the early 21st century (Garfield, 2006). In reality, these are just different branding exercises for the same basic process – quantifying the outputs of researchers with the intention of being able to rank them simply against each other. However, at the core of these processes is an uncomfortable truth. We value these metrics because they can be measured, but they do not necessarily measure that which is valuable. The effect is to distort the landscape – we favour metric-based measurement in an environment where academic critical judgement is far more important (Collini, 2012). It's easy to assess whether a corporation is doing well, in the short term at least, since their outputs can be ranked in terms of production, contribution to GDP, or simple profitability. In knowledge economies, such as academia, the true value of work is far harder to ascertain in the long term and almost impossible to effectively assess in the short term. So we are forced to consider inferior proxies such as journal impact factor (Seglen, 1997) or a scholar's H-Index (Bornmann & Daniel, 2005). There exist other variations and ways of ranking output, but they all suffer from the same basic issue. They are valued because of their convenience and susceptibility to algorithmic analysis rather than because what they represent has true value. They may, in most cases, serve as a proxy for assessing quality but that proxy is usually remote from its destination indeed.

This quantification in higher education does not stop at the gates of a perceived ivory tower. It now extends, at least within the United Kingdom, to the satisfaction of students as expressed on the National Student Survey (NSS). This exerts its pressure primarily by its inclusion as a ranking measure in the university league tables produced annually by several newspapers. Despite long-standing questions over the value of such tables in real terms (Bowden, 2000; Eccles, 2002; Soh, 2011; Turner, 2005), the annual publication of these has become increasingly important events in the academic calendar. As a consequence, they are increasingly influential on decision-making in higher education. Vice chancellors scan the tables with satisfaction or concern, seeking to find the positive news in every bad result and the marketing potential of every success. League tables are informed by university's own measures, but also increasingly by the results from the NSS and REF. Each table weights various factors differently and may include factors omitted by others. This results in a byzantine labyrinth of incompatible ranking measures in which a university may rise dramatically one month in one table and drop dramatically the next month in another. It also gives rise to the preposterous posturing of institutions mining the data for positive nuggets to put on promotional literature. 'Best university in Scotland', 'Best new university in Scotland', 'Best university for staff–student ratio in the north of England', 'First for infrastructure spend'. The outcome of league tables is largely a branding exercise, looking to turn lead into gold through the alchemy of propaganda.

Given the importance of these quantification processes, we should not be surprised that gaming of these systems is endemic throughout the sector. To inflate the impact factor of their publications, journal editors sometimes require new papers to have citations to previous articles within that journal. For those looking to be a little more circumvent in their gaming, they may strike up reciprocal arrangements with other journal editors (Peters & Marsh, 2009; Teodorescu & Andrei, 2014) – you cite my back and I'll cite yours. Sometimes this arrangement extends to circles of journals inflating the impact factor of each other, or more complex webs of obligation and reciprocation. With sufficient diligence, a conspiracy of citation may be extremely difficult to detect.

Individual academics looking to boost both their citation counts and H-indexes may adopt the same techniques on a smaller scale. Academics active in a similar field and with similar research interests may make informal arrangements to cite each other as often as possible, with the aim of making their output look more impressive in an algorithmic analysis. Here, it is almost impossible to effectively distinguish between honest scholarship and self-interested reciprocal citation inflation. A sneaky extra reference can be inserted into a paper with little difficulty by simply mentioning ongoing related work in the background section of the paper. Sometimes it is more blatant (and intentionally satirical), as with the Academia Obscura Kickstarter which offered to cite a paper of the backer's choosing to blatantly inflate an H-index.

Similarly, difficult to effectively delineate is the practice of self-citation. Academics must place their new work in its scholarly context, which will often include reference to their own previous work or to work that is not directly linked but remains relevant. As with citing a colleague, a reference to one's existing work can often be inserted unobtrusively into the text and have it make perfect academic sense even if the paper had no inherent need to reference it (Shema et al., 2012). In this way, particularly prolific authors may find their citation counts and H-Index increase beyond what is truly reflective of the impact of the work.

Again, it is difficult to separate responsible and effective academic practice here from somewhat more pragmatic gamification. Conscious effort on the part of authors is often required to make sure that their own personal citations are not excessive within a particular paper. If we cite within this paragraph Heron et al. (2013) and Heron (2013), then it is actually part of making an argument about the ease at which self-citation can inflate academic H-indexes, and not strictly speaking a cynical attempt to game an inherently gameable system. However, the fact remains that at least to an algorithm, the citation count for those papers increased by one, and cynically, it increased for those of our papers that could most do with an extra citation to increase an H-Index another point. One might argue that this would be difficult to do in most papers where a meta-discussion about self-citation is not contained within – however, we have works on the ethics of plagiarism (Heron & Belford, 2016), teaching ethics through case studies (Heron & Belford, 2015) and ethics in video games (Heron & Belford, 2020) that would be perfectly consistent to cite within the context of a book on issues of professional conduct and computer ethics. This could easily be done with only the slightest unobtrusive detour in the narrative of the literature review, which, the observant reader may notice, is exactly what we've done.

That may seem thoroughly abuseable, and all we are doing is providing evidence for the prosecution as to our own professional impiety. That belief though lacks full context.

For that, we would like to remind the reader that this book is an expanded and revised version of the Scandal in Academia as was serialised in the ACM's SIGCAS Computers and Society journey. It is based in part on papers already published and already in the academic record. The above paragraph has now inflated the citation count of these papers **twice.** One should be wary of bibliometrics – as the management adage goes, once a quality measure becomes a target, it ceases to measure quality.

Academics, as individuals, are all aware of these issues. Yet we still cleave to the value of bibliometry within a broad community of practice. H-Indexes, impact factors, and citation counts are all commonly used to differentiate ourselves from colleagues, and for promotion committees to justify preferment. We value that which can be measured, even if what we measure has no value. It's a cyclical structure – the various calculations for impact factor are used to generate the value of papers for REF analysis, which then encourages academics to publish in those journals to enhance their own REF rankings. Those in turn are used to inform the perceived value of journals to publishing academics. It is not fair to say that the REF does not identify high-quality, high-impact research – gaming these statistics can work for middling ranks, but not for those at the highest thresholds. However, by definition most work assessed by processes like these will lie closer to the median than to the maximum.

Similarly with the quantifying of student satisfaction, we find here an inherently gameable system. It is not to say those that do especially well in the exercise are not objectively better than those that do poorly (Sabri, 2013). However, within bands of the results what we tend to see is the institute that was best at gaming the result attains the highest results. Satisfaction in all measures within the NSS tends to band in the 80%–100% region, and thus every tiny element of extra approval can have a disproportionate impact on rankings (Frost, 2021). Thus, we encounter a broad spectrum of attempts to game the system – some largely well meaning, some more dubious (Cheng & Marsh, 2010).

At the more ethically pure end of the spectrum, we find institutions trying to encourage a certain minimum participation in the exercise, full in the knowledge that the greater the participation rate (to a limit) the more likely that negative feedback will be smoothed over. As Harry S Truman famously remarked upon receiving yet another mean-spirited piece of correspondence about his presidency, 'Why is it only sons of bitches know how to lick a stamp?'.

Encouraging full participation means that those hypothetical sons of bitches have their views moderated by the more reasonable viewpoints of those that did not feel inherently compelled to participate in the process. At the other end, we find blatant exhortations for positive results, such as lecturers at an unnamed university that were recorded as saying the following[3]:

> If (university redacted) comes down the bottom, the bottom line is that nobody is going to want to employ you because they'll think your degree is shit.

Guidance issued by HEFCE (partly responsible for administering the study) is strict about what institutions may say to students regarding guidance on completion. That does not stop certain persistent myths being promulgated. Some institutions tell students that 'Neutrality is analysed as disagreement in the stats, so if you're going to be neutral just be positive' – this is both incorrect and so prevalent that HEFCE has issued specific feedback regarding that particular claim. It is important to note here that HEFCE and the universities themselves deny there is any endemic problem with manipulation of the system. That is in part an expression of the core problem. Many of the ways in which the system can be gamed are invisible to any analysis. Universities have been known to time high-profile positive announcements so that they coincide with key NSS dates, ply students completing the test with free drinks and pizza, and hold back bad grades until after the NSS surveys are completed. None of this need be especially obviously a case of manipulating the system, but it's all likely to have a significant impact. After all, if the severity of custodial sentences modulates with a judge's proximity to lunchtime (Danziger et al., 2011) it's not possible to discount the powerful effect that a temporary attitude has on quantifying that which is inherently subjective.

None of this is to say that quantification, in and of itself, is a bad thing – it is only right and proper that publicly funded institutions such as universities give an account of what they do for the public that funds them. Private organisations live within a far more fraught culture of quantification than university academics, and customer satisfaction surveys are a common measure in industry. The authors of this book are in no way trying to say that universities should be immune from oversight or monitoring. However, it is important that we place quantification into its appropriate context. It is a flawed process prone to gamification and flattening of nuance, and inherently unsuited to measuring that which cannot reasonably be measured. Assessing quantified data must be done with that in mind.

Within the Scandal in Academia case study itself, we see the end-point to which this chase for research and ranking can lead. We don't see how the system works and how the demands from the university that may seem reasonable to an outsider are impossible to guarantee given the structural inequalities of the system. It's not enough to submit a grant for a million pounds – it's necessary to write many grants for a million pounds, with the hope that one of them will be funded for long enough to meet income demands. Few people would find much solace in a job requirement that included the deliverable 'regularly win the lottery'. And yet, that is what our current culture of fraught and unreliable research funding requires. Given that, perhaps we might see Blackbriar's actions in a different light. Or indeed, perhaps not.

CIVIC ENGAGEMENT, CULTURAL IMPACT, AND THE UNIVERSITY

Communicating the specifics of what a university is actually for is surprisingly difficult (Collini, 2012). One of the common things that tends to be generally assumed of the modern institution is some degree of civic engagement. Universities are expected to be able to provide at least some insight and illumination on matters important to society even when the investigation of this may not be economically feasible. Public outreach is considered to be part of this, communicating what is state of the art in research or scholarship to those outside of the lecture halls of the institution. As an example of the form this might take, the authors of this book have been communicating accessibility research in tabletop gaming to the public via the Meeple Like Us blog [4]. This takes issues that would otherwise be confined to an academic journal paper (Heron et al., 2018a, 2018b) and casts them in a way that is tractable to the general public.

In modern Higher Education PLC, this kind of activity might be considered a form of being a good corporate citizen or more formally a part of the mandate of the institution's charter. For some, it may be a self-conscious attempt to counter accusations that academics live in an ivory tower.

More positively it can be viewed as an important part of meeting the promise of education as a public good and as a fulfilment of implied societal obligations (Ecklund et al., 2012). For others, it may be an attempt to build their own academic profiles, short-circuit traditional publishing loops, and guide cultural discussions.

Acting in the public good was certainly the role that Professor Joanne Clements within the Scandal believed she was filling. She believed that the hard-nosed economic success of other departments should subsidise her own less profitable but culturally important role. Again, this is an issue where reasonable people may disagree. It can be hard to see the public good in academic discourse which is often seemingly explicitly engineered to prohibit real-world impact. We can see this in the often insular critical discussions around obscure texts that dominate certain literature departments.

Much of this scholarly discourse is explicitly inaccessible, employing a jargon and a presumptive body of knowledge that prohibits any outsider meaningfully engaging with the work. The perception for many is that these fields are cliquish and ultimately ancillary. Their academic publications are perceived to be aimed at only a handful of others interested in the esoteric intricacies of a topic which the rest of the world is not invited to enjoy.

Public grumbles of those that feel as if they should feel some benefit from their taxes are joined by the grumbles of those that feel they are subsidising the self-indulgent hobbies of others. However, it is rarely the case that these academic discourses are truly isolated from the real world, and will often come to inform future scholarship when rarefied deliberation of isolated academics filters down into the more accessible critical literature. From there, it eventually makes its way down closer to Earth in the curricula of university and high school English departments where it may, optimistically, go on to inspire or energise the wider public as part of a formal education programme. Part of the benefit of a university is that it can devote some of its focus to long-term impact in a softer, subtler way. As Collini (2012) notes in his Diary of a Don:

> ++++ Home to lick wounds. Console self that the real thing, as all scholars know, is not the immediate reception of one's work, but whether it goes on to play its part in stirring the thinking of its readers, some of whom may use it to help make their own book a little bit better. When that happens, then we really do have a

> news headline to match all the fanfare about those 'discoveries' by our scientific colleagues: 'Cambridge scholar cited in footnote to obscure monograph'. Hold the front page.

This is in part what the block funding settlements of the various research funding councils aim to do – to offer a certain proportion of funding aimed at supporting basic, or blue sky, research without the need to explicitly tie it to concrete goals and deliverables. However, as we can see above with our discussion on the REF, high-impact research leads to high-impact publications, which leads to high REF scores, which leads to greater funding. The constant seduction is to spend the money where the results will be most obvious so as to generate greater funds for the next research exercise. Research into softer disciplines with less opportunity to make headlines or inform popular cultural discourse offers far more nebulous returns. Similarly, working with local groups or charities offers less chance at sustainable long-term academic returns, even if the need in the community may be greater and the ability of the university to influence society in a modest way is more significant. It becomes a balancing act between allowing for exploration of issues in the public interest and enhancing the future sustainability of the institute in an increasingly uncertain and competitive funding climate. Corporations are accountable to their shareholders, but it remains uncertain to whom universities should ultimately report.

DATA, INFORMATION, KNOWLEDGE, AND WISDOM

We've already spoken about publication pressure, where at high-performing institutions an academic may be expected to produce between two and four papers – of suitable prestige – every year. Sometimes this gets done by micro-slicing research down into its minimum viable publication form – work that would have once been reported in a single paper becomes two, three, or four smaller publications. Big ideas get watered down to fit the expected tempo of an academic career. Peter Higgs, recipient of the Nobel Prize for Science in 2013, has said that he wouldn't make the cut for modern academia. After his first major contribution to the literature on physics in 1964, he published fewer than ten subsequent papers. Speaking in the Guardian, he said that had he not been nominated for a Nobel prize in 1980, he almost certainly would have been sacked from

Edinburgh University. His subsequent security at the institution was a kind of long-term gamble on him eventually receiving the award [5].

> Higgs said he became "an embarrassment to the department when they did research assessment exercises". A message would go around the department saying: "Please give a list of your recent publications." Higgs said:" I would send back a statement: 'None.'

Few of us in academia are likely to receive a Nobel prize, or an award of equivalent stature. Higgs added:

> It's difficult to imagine how I would ever have enough peace and quiet in the present sort of climate to do what I did in 1964.

In this, we can see an echo of what Clement reported to the Dunglen Chronicle. Once upon a time, the job of an academic was to have big ideas and spend a lifetime pursuing the implications. It was a job that was profoundly based upon the building – and inculcating in the student body – of **knowledge**. There is a somewhat loose hierarchy that we can apply here. At the lowest tier of understanding is **data**, which is a raw collection of numbers and letters brought together into a format for easy conversion, processing, or transportation. A list of numbers, or a series of alphanumeric characters. Measurements or observations fall into this bracket. We hope that data is made up of facts, but by now we know better.

Data can then be processed into **information**, which is data that conveys context. A list of alphanumeric characters is data. It becomes information when it is contextualised as a list of postcodes. A set of real numbers is data until we find out it reflects measures of atmospheric pressure. Data lacks organisation, because the act of organising requires understanding of context. Information on the other hand is organised and amenable to analysis. We refine, through the application of context, data into information.

Knowledge is the higher form of information, in which it is given meaning and depth. Knowledge allows for the comparison of information against other kinds of information, and for the creation of **new** information that derives from the old. Knowledge is the shape we give to information in order for us to make judgements and deduce impact. Knowledge tells us that a tomato is a fruit, or a tomato is a vegetable, depending on its context – whether we are looking at it in a botanical frame or a culinary one.

We might even argue that there is a tier above that – **wisdom**. As the old joke goes, knowledge tells us that a tomato is a fruit. Wisdom tells us not to put one in a fruit salad. Wisdom is a refined form of knowledge in that it is built from perspective, sound judgement, and deeper analytical insights derived over time from ongoing consideration of the evidence. The most valuable new contributions to our collective understanding come from people using their wisdom to generate thoughts that others can't.

It would be hard to argue that modern research does not produce spectacular results on occasion. However, these results come with great costs. The production of a paper is a collaborative effort, requiring authorial input, editorial judgement, and reviewer critique. The sheer number of publications that get submitted to any conference results in a tsunami of need set against a light wave of availability. We will paraphrase a friend of ours – a long-serving associate chair for several high-profile conferences in his field. When talking about finding qualified reviewers for specialist papers:

> It's just a scramble to find someone that might have a slight interest in the work, or even knows something about a different anagram of the field to be reviewed.

What is the purpose then of a university academic? Is it to annually produce two papers of information – or perhaps at best knowledge – and thus contribute to the over-production of academic papers across their field? Or is it to allow knowledge to evolve naturally over time into wisdom? Clement talks about academia as a 'write only process', where success is defined as being able get a paper out at a high-impact conference. Having attended these in the past, the authors of this book can say people most often seem to spend much more time thinking about what they're going to present than they do listening to the presentations of others.

Berg and Seeber (2016) argue convincingly that the speed of academia is its largest problem – that we need to give big ideas a place where they can mature, and not worry so much about our bibliometrics. Unfortunately, at least for now, this is wisdom that is really only actionable by those with stable careers and reliable tenure. It took Peter Higgs and his team around 50 years to get the proper recognition of how his big idea changed the world. A recruitment and promotion strategy that takes into account only the previous output of the last five years seems unlikely to incentivise developing that kind of wisdom. That is especially true given how many papers are never read, and how many more are never cited. It has become a Sisyphean task for an academic to keep up with the output even within relatively narrow fields of specialism. There is, as Altbach and De Wit (2019) argue – too much research being published:

> There is too much pressure on top journals, there are too many books and articles of marginal quality, predatory journals are on the rise, and there is tremendous pressure on academics worldwide to publish.

Still, in an increasingly quantified world, where universities live and die by their bibliometrics, it is hard to know what the alternative should be, particularly given that universities do not decide their priorities in a vacuum. Quantification of unquantifiable research value is increasingly demanded by funding bodies and the governments that supply them with their money. It is not unreasonable that they should see some returns on that investment of public money and that requires reporting on results. Clement believes the university – as a public institution – has lost its way. Taylor believes that it is a dangerous delusion to cling to the ivory tower of the past.

For the part of the authors of this book, we believe part of the problem is in how academic work is contextualised. The word 'research' implies a process that establishes new facts (generates data) and reaches new conclusions (outputs knowledge). Thus, when we think of universities as research institutions, we bring those assumptions with us when we grade them on their outputs. Perhaps a subtle rebranding is in order to show that there is still room for a more gentle form of academic output that comes from engaging with the act of **scholarship**. Instead of academics thinking of themselves as researchers, we should perhaps be more prepared to think of ourselves as scholars – that our end goal is of somewhat grander ambition than to simply say something new in a form that contributes cleanly to our career path. As soon as someone works out how to square that with the demands of governmental oversight, that is.

But still, it's hard to say that this would be inherently preferable over the current system, in which the speed at which knowledge is produced is perhaps one of the great achievements of the university sector. The question is what we believe is most valuable to society, and that's something about which everyone may have differing opinions.

NOTES

1 https://www.sfc.ac.uk/publications-statistics/announcements/2022/SFCAN152022.aspx
2 https://www.timeshighereducation.com/news/imperial-college-professor-stefan-grimm-was-given-grant-income-target/2017369.article
3 http://news.bbc.co.uk/1/hi/education/7399059.stm
4 https://meeplelikeus.co.uk
5 https://www.theguardian.com/science/2013/dec/06/peter-higgs-boson-academic-system#:~:text=Higgs%20said%20he%20became%20%22an,'%20%22

REFERENCES

Altbach, P. G., & De Wit, H. (2019). Too much academic research is being published. *International Higher Education*, (96), 2–3.

Arnott, M., & Menter, I. (2007). The same but different? Post-devolution regulation and control in education in Scotland and England. *European Educational Research Journal*, *6*(3), 250–265.

Berg, M., & Seeber, B. K. (2016). *The slow professor: Challenging the culture of speed in the academy*. University of Toronto Press.

Bornmann, L., & Daniel, H.-D. (2005). Does the h-index for ranking of scientists really work? *Scientometrics*, *65*(3), 391–392.

Bowden, R. (2000). Fantasy higher education: University and college league tables. *Quality in Higher Education*, *6*(1), 41–60.

Browne, J. (2010). Securing a sustainable future for higher education [Browne report]. https://assets.publishing.service.gov.uk/media/5a7f289540f0b62305b856fc/bis-10-1208-securing-sustainable-higher-education-browne-report.pdf (accessed 30 October 2015).

Cheng, J. H., & Marsh, H. W. (2010). National student survey: Are differences between universities and courses reliable and meaningful? *Oxford Review of Education*, *36*(6), 693–712.

Collini, S. (2012). *What are universities for?* Penguin.

Danziger, S., Levav, J., & Avnaim-Pesso, L. (2011). Extraneous factors in judicial decisions. *Proceedings of the National Academy of Sciences*, *108*(17), 6889–6892.

Eccles, C. (2002). The use of university rankings in the United Kingdom. *Higher Education in Europe*, *27*(4), 423–432.

Ecklund, E. H., James, S. A., & Lincoln, A. E. (2012). How academic biologists and physicists view science outreach. *PLoS One*, *7*(5), e36240.

Frost, N. (2021). *The myth of measurement: Inspection, audit, targets and the public sector* (pp. 1–100). Leeds Beckett University.

Garfield, E. (2006). The history and meaning of the journal impact factor. *JAMA*, *295*(1), 90–93.

Giske, J. (2008). Benefitting from bibliometry. *Ethics in Science and Environmental Politics*, *8*(1), 79–81.

Heron, M. (2013). "Likely to be eaten by a Grue"—The relevance of text games in the modern era. *The Computer Games Journal*, *2*(1), 55–67.

Heron, M. J., & Belford, P. H. (2015). A practitioner reflection on teaching computer ethics with case studies and psychology. *Brookes eJournal of Learning and Teaching*, *7*(1).

Heron, M. J., & Belford, P. (2016). Musings on misconduct: A practitioner reflection on the ethical investigation of plagiarism within programming modules. *ACM SIGCAS Computers and Society*, *45*(3), 438–444.

Heron, M. J., & Belford, P. H. (2020). Do you feel like a hero yet? *Journal of Games Criticism*, *1*(2).

Heron, M. J., Belford, P. H., Reid, H., & Crabb, M. (2018a). Eighteen months of meeple like us: An exploration into the state of board game accessibility. *The Computer Games Journal*, *7*(2), 75–95.

Heron, M. J., Belford, P. H., Reid, H., & Crabb, M. (2018b). Meeple centred design: A heuristic toolkit for evaluating the accessibility of tabletop games. *The Computer Games Journal*, *7*(2), 97–114.

Heron, M., Hanson, V. L., & Ricketts, I. W. (2013). Access: A technical framework for adaptive accessibility support. In *Proceedings of the 5th ACM SIGCHI symposium on Engineering interactive computing systems (EICS '13)* (pp. 33–42). Association for Computing Machinery, New York, NY. https://doi.org/10.1145/2494603.2480316.

Joyner, J. (2014). Unprofitable professors getting fired. *Outside the Beltway*. http://www.outsidethebeltway.com/unprofitable-professors-getting-fired/.

Martin, B. R. (2011). The research excellence framework and the 'impact agenda': Are we creating a Frankenstein monster? *Research Evaluation*, *20*(3), 247–254.

Merton, R. K. (1968). The Matthew effect in science: The reward and communication systems of science are considered. *Science*, *159*(3810), 56–63.

Mooney, G., & Poole, L. (2004). 'A land of milk and honey'? Social policy in Scotland after devolution. *Critical Social Policy*, *24*(4), 458–483.

Peters, J., & Marsh, R. (2009). Rate my research dot com: Measuring what we value, and valuing what we measure. *Management Decision*, *47*(9),1452–1457.

Sabri, D. (2013). Student evaluations of teaching as 'fact-totems': The case of the UK National Student Survey. *Sociological Research Online*, *18*(4), 148–157.

Scales, P. A. (1976). Citation analyses as indicators of the use of serials: A comparison of ranked title lists produced by citation counting and from use data. *Journal of Documentation*, *32*(1), 17–25.

Seglen, P. O. (1997). Why the impact factor of journals should not be used for evaluating research. *BMJ*, *314*(7079), 497.

Shema, H., Bar-Ilan, J., & Thelwall, M. (2012). Self-citation of bloggers in the science blogosphere. *Science and the Internet*, 183–192.

Soh, K. C. (2011). Don't read university rankings like reading football league tables: Taking a close look at the indicators. *Higher Education Review*, *44*(1), 15–29.

Teodorescu, D., & Andrei, T. (2014). An examination of "citation circles" for social sciences journals in eastern European countries. *Scientometrics*, *99*(2), 209–231.

Turner, D. (2005). Benchmarking in universities: League tables revisited. *Oxford Review of Education*, *31*(3), 353–371.

CHAPTER 8

BrokenBriar Affair Heating Up

Lawyers Involved

NEWSPAPER ARTICLE

An exclusive report by Jack McKracken for the Dunglen Chronicle

The suspended postgraduate students at the centre of what has become known as the 'BrokenBriar Affair' have today turned up the heat on the University by enlisting the services of respected local lawyer Karan Chandra. Chandra, a prominent citizen in Dunglen, has taken on the case pro-bono and has stated that any settlement fees that would go to him will instead be awarded to the University of Dunglen Hardship Fund. 'I believe this is a matter of human rights, not legal point-scoring', he said. 'As such, it's my moral duty to lend my assistance'.

Speaking today on behalf of Duncan and McAlpine, Chandra said, 'This is one of the most egregious cases of institutional bullying I have seen in my career. My clients are victims in this incident and they have been treated by the university as if they are guilty of some kind of offence. Their budding professional reputations have been destroyed by the way in which the university has sought to take the heat off of its own organisational malfeasance. They will no longer be talking to the press. If you have any enquiries, they should now be directed to me'.

DOI: 10.1201/9781003426172-8

The University of Dunglen has become increasingly unresponsive over the scandal in recent days, prompting others to investigate legal options to compel disclosure. The university's legal department has issued a statement saying that neither the university nor the individuals employed within the university can comment on what is now an ongoing legal matter.

However, an insider within the University Court spoke to the Chronicle saying, 'The shit has hit the fan, basically. All the allegations are true – the data set is riddled with invented data. We got hold of the original figures as part of a joint investigation with the oil companies, and basically now it's a case of damage limitation. Problem is, you can't limit this kind of damage because it's already gone beyond what we can manage internally. The Blackbriar Algorithm has been overstating the amount of extractable resources by a factor of five because it's been calibrated against what is essentially wishful thinking. It's not completely fraudulent, just always giving the best results rather than the likely results'.

When asked about the case of Mrs McAlpine and Mr Duncan, our source said, 'They've been treated like sacrificial lambs. The principal hoped that a blood offering to the press might suffice and that it would all blow over. His plan was to make it look like they'd just misreported the data. John would be cleared, we'd quietly retire the data set, our contracts would keep ticking over and we'd find a way to get rid of John when it wouldn't be so damaging to the university. Obviously that plan has backfired dramatically and now we're basically in siege mode'.

It is reported that Chandra has already begun to construct the case required for an academic appeal and will be insisting upon a full public vindication of his clients. Parallel to this, he is instituting legal proceedings to claim for reputational damages, which are likely to be significant, and for full intellectual property for all work done by the students to be transferred to them from the university.

'Let's be honest, they can't finish their doctorates at the University of Dunglen now. The best we can hope for is full exoneration for them. We also want to take what research they can salvage to another institution that won't treat them quite so disgracefully. Whatever happens, they'll suffer from this stigma for their entire careers and so I am also seeking reputational damages commensurate with the scale of their betrayal'.

Professor John Blackbriar has also been seen in the university in recent weeks. The professor who is the epicentre of the crisis has been in hiding since the scandal broke, refusing to respond to press queries or answer phone calls. He was seen leaving his office in the early hours of the morning after a secret, emergency meeting of the University Senate had been called. The Chronicle understands that the professor too has retained legal counsel. No official statement has been forthcoming from the professor's legal team but we are assured one will be released 'in due course'.

The revelations last week regarding the university's culture of research have had significant ramifications within the postgraduate and postdoctoral body of researchers. The extent to which their environment had been consciously engineered by the vice chancellor has come as a shock to many who believed it was simply a consequence of professors seeking academic excellence. One postdoctoral fellow spoke to us, saying, 'I feel betrayed. It's one thing to work these hours for the love of producing knowledge. It's another because your boss is grinding you into the ground to protect their own position'.

Another added, 'I feel used. That minute spoke of us in the same way as it spoke of lab equipment – we're just tools to be used, not people to be respected. My career prospects are not even an after-thought – it's just a case of getting the most they can out of us and then leaving us mentally burnt to the ground'.

Within the sector, the postdoctoral researchers have numerous issues with career stability and often lack any significant leverage to improve their day-to-day positions. Research projects normally come with an expiry date and a fellowship may only exist for two or three years. At the end of this period, doctoral fellows are expected to either move to another university where grant funding may be available or find shelter in the safe harbour of a permanent academic position such as lecturer. However, it has been estimated that there is only one opening for every ten individuals for the latter.

Dr. Jake Dymock, employment consultant to the University of Alba, told us 'Universities have for years increased the number of doctoral researchers they have generated, way beyond the ability of the academic sector to absorb with careers. A doctorate isn't like a "better degree" – it opens a lot of doors, but they're almost all in academia. It closes a lot more – you specialise so much in a doctorate that you lose out on the ability to train career-relevant skills in areas outside your topic. So, imagine ten people running for every single open door – that's academia. The majority who don't get through a door are left with impressive but short-term research careers and yet no reward for the work they put in, and no real options except to look for another short-term bridging contract. That is – if they even want to stay working in academia. Increasingly, people are looking for escape ramps to less exploitative career paths'.

'It's a hard egg to crack', one postdoc tells us. 'On the one hand, grants need to come in to employ people like me. That needs results and lots of them. On the other, we're supposed to be developing our careers too. We're not supposed to be cogs that you can clip into a machine and then discard when you're done with them, and yet that's how many of us are treated'.

As the scandal-beset University tries to fight off legal challenges and a growing discontentment within its research base, undergraduate students have sided with the university's critics. There have already been several student protests on campus, including numerous sit-ins and walk-outs. 'We know how hard some of these guys work' said a third-year sociology student. 'I don't want to graduate into a world where I'm just a number on a spreadsheet. We need to treat each other better, and the fight for that begins here'.

The growing scandal at the university is not confined to the pages of the Dunglen Chronicle. The larger social media platforms are increasingly becoming a major source of content regarding the situation. Video essays, vlogs, reaction videos, and more are becoming increasingly important in the battleground of this public relations war.

THE DISPOSABLE ACADEMIC

The plight of those working within highly quantified and structurally unequal systems is only part of the case study – we must also look at the way in which the institution has treated those students it suspended in the early days of the crisis. We have not heard much from them, likely because the wheels of litigation turn slowly if they turn at all. In this section of the case study, we find that in the background Sharon McAlpine and James Duncan have been taking legal advice on their suspension. Blackbriar too has retained legal counsel, reflecting the highly charged climate of litigation and counter-litigation that is becoming the norm in academic and professional contexts. For Blackbriar, it's not surprising – disclosures from 'an insider' within the University Court suggests things are looking bad for the embattled professor. The procedures and policies of modern universities may be drafted in consultation with lawyers, but they are most often enacted by academics and bureaucrats with only passing familiarity with the strictness of legal interpretation. As such, few universities are very comfortable with the idea of lawyers being brought to bear in issues where academic, managerial, or professional judgement has been a major factor. For issues of workplace grievance within the United Kingdom, arbitration bodies such as the Advisory, Conciliation and Arbitration Service (ACAS) serve to insulate employers from the courts. However, for other issues there are few recourses except to call in the lawyers. Students in particular have become increasingly litigious in recent years although concrete figures are hard to come by – universities as a whole are not keen on publicising these confrontations, partially out of the fear of giving anyone funny ideas. However, anecdotal data suggests a growing influence of legal pressure with regard to rendered judgements in cases of plagiarism; academic misconduct; issues of accessibility to resources; and grade profiles. Some law offices even explicitly offer academic services aimed at enforcing breach of conduct charges against universities that are delaying, suspending, or cancelling student enrolment.

The stakes are high, especially in the English and Welsh universities where a student may be paying as much as £9250 a year for tuition. They become astronomical when we consider American universities where annual costs can range from $12,000 in a public two-year college to $50,000 for a private university education. The potential rewards for cheating are directly informed by the perceived value of the qualification weighed against the perceived difficulty of achieving a good grade. Cases of academic misconduct are increasingly common across all institutes, and as pressure is brought to bear on internal procedures it becomes increasingly unlikely that all will be conducted with the rigour necessary to prevent legal challenge (Heron & Belford, 2016).

Complaints against universities have become commonplace – 2022 was a record year for complaints within the UK, with a total of 2,850 being received by the Office of the Independent Adjudicator for Higher Education.[1] 2022 was the fourth year in a row in which the number of complaints in a year had risen. Almost 40% of these complaints related to grades or progress. After an investment of as much as £27,750 in tuition fees (discounting the associated living expenses) for a chance at future professional success, a roll of the dice on a grievance procedure seems almost common sense. For those with the financial backing to hire legal representation, it's also possible to capitalise on the university's natural reluctance to wage costly legal battles over issues of academic judgement.

We haven't yet had cause to fully explore this in relation to our case study, but later articles bring this and other issues more prominently to the fore. For now, it is sufficient to acknowledge the risk that litigation brings to the University of Dunglen and the academics working within.

We will find now our students are increasingly willing to disclose some of what might be considered the day-to-day operational intelligence of the institute. It's not quite whistle-blowing, but it does veer into the grey area between professional indiscretion and reporting on institutional misconduct. Chandra notes himself 'their budding professional reputations have been destroyed', and as discussed in previous chapters we now live in a culture where perceived transgressions have a toxic half-life measured in decades. Their careers are likely in tatters – their proximity to a public scandal will taint their lives from this point onwards regardless of what else happens. Social media and those that have charged themselves with the role of aggressively policing tone and behaviour will ensure that whenever anyone googles them in future it's the scandal they'll find. They will need to live with that fallout and find a way to rebuild (Ronson, 2016). McAlpine and Duncan are not entirely reliable narrators in this – having been mistreated by the university they are unlikely to be entirely objective observers of events. That however does not mean we can simply discount their account particularly when it is echoed and corroborated by the anonymous testimony of other students in the newspaper.

Much of it rings true for anyone who has experience with the research postgraduate programmes in numerous universities. Research students are often insulated from the day-to-day politics of an institute, but subject to the consequences of diktats from on high. The high-pressure environment in which our students were working may have been shaped in part by the professional priorities of John Blackbriar. Those priorities were shaped in turn by the principal. They talk of being offended by the way in which they were seen – lab resources and assets, cogs in a machine designed to create impact and citation and income. They were shocked by the callousness with which their well-being and 'work-life balance' were disregarded with the blithe directive to 'increase the amount of useful value we get out of these resources'. Regardless of how many hours more than is healthy they may have been putting into their budding careers, they were brushed off as being 'underutilised'.

Abstract distance from people makes it easy to treat individuals as mere entries on a spreadsheet. Postdocs spoke of being 'betrayed', knowing now that they were being overworked to protect the professors, rather than out of an inherent desire to ensure their careers were being forged in the right way. For many, this may have been their first real lesson in the often callous calculus at the heart of modern industrial relations – you have little inherent value to the organisation other than that which is produced by your labour.

Part of the problem here is that ultimately it's true – the culture of research is such that those individuals that find themselves adrift between permanent contracts often are seen, at least subconsciously, as resources to be worked for maximum benefit for the research project.

These projects may only last a few years, at which point the 'resource' will either need to apply elsewhere for likely another temporary placement or seek funding to sustain their career at their current location. More and more important and skilled colleagues simply abandon their pursuit of an academic career entirely (Shelton et al., 2001) and move into the private sector (Fritsch & Krabel, 2012). Progression from temporary contracts on to permanent contracts is vanishingly rare – remember that it's estimated that there is only one opening for every ten applicants. The effect is endemic and disheartening – a culture of treating valuable, dedicated, and highly qualified professionals as disposable assets. For women, the situation is especially troublesome given societal expectations regarding motherhood and the incompatibilities these often have with career progression. We cannot discount these intersectional issues.

The growing casualisation of higher education has a corrosive effect to both individual careers and the long-term conduct of research. This is a consequence of the way in which we have quantified the academic experience. Research funds that are transient, unreliable, and increasingly tied to 'impact' are simply not sustainable enough to allow universities to hire researchers on a more permanent basis. This coupled to an overproduction of postdoctoral applicants for research positions that drives down the need to treat any one individual as valuable. That's not to say everyone is mistreated and maligned during their early academic career, or even that it's true for the majority. However, the pressures increase in line with the prestige of the department in which a researcher may find themselves (Niles et al., 2020). The competition, for one thing, is notably fiercer.

The often misguided expectation from research fellows is that accumulated research prestige can eventually be parlayed into a permanent academic position. In such environments, a model of 'obsessive workaholism' seems to dominate discourse, with academics reporting that they regularly spend more than 60 hours a week working, with the concomitant impact on stress and health levels. We may have genuine cause to question the accuracy of these self-reported measures and how productive those 60 hours could be. However, it's certainly the case that researchers often feel the need to be seen to be working. As regressive as it is, persistent long hours in academia is still perceived as a symbol of dedication rather than inefficiency or a symptom of an unhealthy work-life balance. Again, this is not unique to academia – software engineering as a discipline is particularly plagued by the culture of the 'crunch' (Kwak & Stoddard,

2004) and in certain cultures it is considered extremely bad corporate etiquette to be the first person to leave for the day. This in turn creates an escalating arms-race of unpaid overtime. With steady employment being increasingly difficult to procure, the power is very much in the hands of the employer and those empowered to enact their policies. This is also not new – an unearthed letter from 1996 from an associate professor to a research assistant sets out a number of thoroughly unreasonable demands at Caltech:

> I have noticed that you have failed to come in to lab on several weekends, and more recently have failed to show up in the evenings. Moreover, in addition to such time off you recently requested some vacation time…

And:

> I receive at least one post-doctoral application each day from the US and around the world. If you are unable to meet the expected work-schedule, I am sure that I can find someone else as an appropriate replacement for this important project.

Such overt and specific threats may not be common these days due to the risk of litigation or censure over social media. The practices remain an unspoken assumption in many high-profile departments. Autonomy may be limited, professional credit may be doled out in miserly quantities, and the ability to influence research directions can be minimal. However, if you are working for a research superstar or are attached to an internationally renowned department, the personal cost may be worth the professional benefits – should they materialise. There is though a growing realisation amongst young researchers that academia is a Ponzi scheme where there is all take and very little if any genuine give. A bad postdoctoral placement places all the power in the hands of the professor – not only is the current position within their gift, but it is their recommendation that will be pivotal in securing the next. Postdocs are left with the choice between 'toughing it out' or 'blowing the whistle'. The latter rarely offers much more than temporary satisfaction.

The minute from the University of Dunglen lays that out in stark terms for the researchers in the case study – some shock, upset, and anger is entirely to be expected. To a certain extent, we make our peace with the context of the work by focusing on its more intangible benefits, such as perceived flexibility and ability to pursue that which is intellectually validating.

Higher education is not yet at the point where the authors of this book actively discourage academically minded young people from pursuing a PhD, but we do strictly limit the parameters under which we'd recommend it as a programme of study. In this, we are not alone – increasingly those with relatively secure positions are telling those trying to enter the ivory tower that the drawbridge was raised against newcomers a long time ago.

WE VALUE WHAT WE CAN MEASURE

A theme across this and the previous article has been related to performance targets – that the feeling associated with the scandal across the community at large has been one of betrayal. One of the informants talks of the difference between doing what they do for the love of the work versus being squeezed for every last ounce of productivity. We've spoken at length about publication expectations and funding targets, but we haven't yet really delved into what this means in practice for other fields – some of this feels very 'academia-specific'. The truth is – every industry has versions of this. All that changes is the form that the performance evaluation takes.

Within academia, it is the flexibility that many people enjoy. The pressures can be high, but the reward is an almost unmatched sense of autonomy. It is not uncommon for senior managers in a university to baulk at this, protesting that academics need to stop thinking of themselves as largely self-employed consultants and more like employees. However, the tradition of academic contracts offers protection from some of these transactional assumptions. Or at least, it used to offer protections. Many contracts make more established academics difficult to dislodge from their position. Academic contracts, and academic norms, often include protections for academic freedom and establish flexibility to teach and research in the way an individual academic feels appropriate. Academic unions tend to be relatively strong, at least in comparison to their national baseline. Academics tend to stay in institutions for a long time, meaning they are expensive to get rid of through redundancy. Academic programmes can be built around an academic's specialised skillset, making replacing them difficult. That's not to say tenured academics are safe from the career uncertainty that is taken for granted in other sectors – just that they can take a somewhat more relaxed approach with regard to the risks associated with routine dismissal.

We've already encountered the case of Stefan Grimm who received an ultimatum regarding his grantsmanship, and reportedly had been told he'd be dismissed if he failed to bring in the expected income. The university has a different view, saying that his dismissal was never under consideration and that internal evaluation had only ever reached the stage of informal performance management. What is said in public and what is said in private may differ – we're not in a position to be able to say which account is the more accurate one. However, even given the informal performance management there is a tone of menace – specifically the setting of an unreasonable expectation (being awarded a grant in 12 months) and with reference to the disciplinary procedure of the university. Given that this is part of the broader terms and conditions under which the professor operated, they serve as a way to set the parameters upon which tenure may be terminated. It is clear from the university bylaws that even the intention to dismiss comes at considerable bureaucratic cost.

That's not a luxury enjoyed by most working professionals. Video game companies routinely lay off highly qualified and experienced staff to boost the bottom line for shareholders. Tech workers often find themselves at the mercy of algorithmic evaluation such as stack ranking (Lopp & Lopp, 2021) where employees are categorised as high, average, or low performance relative to their colleagues. At some companies, the bottom 10% of the stack may find themselves laid off as a matter of annual or intermittent downsizing (Van de Poll & Kroese, 2022). Depending on where you are in the world, you might find yourself subject to what is known in the United States as 'at will employment', in which any legally appropriate reason may be used to fire someone whenever and without warning.

The stability of a career depends on many factors, and geographical location is one of them. Sweden has a law (The Swedish Employment Protection Act) which enshrines the doctrine of 'last in, first out' when redundancies must be made. Older workers in this have more protection than younger workers. Union membership in the Scandinavian countries too tends to be very high, with decisions regarding working practices and culture being decided in negotiation between management and the workers. In the United States, you may on the other hand find that even raising the topic of starting a union chapter results in your position becoming untenable. In many cases, being fired for attempting to unionise is unlawful, but managers can often find a legal reason why a troublesome person should be made to vacate their job.

Performance management serves an important and useful role in any organisation. It allows for employees and employers to reach a common understanding as to how well someone is doing. It allows for alignment of workers to the overall goals of the organisation. It can motivate high performance, if the review aligns with incentive structures. It can help employees become aware of contributions that they themselves would not have claimed. It allows for problems to be identified before they become disciplinary issues, and can identify opportunities for employees to grow into other roles within the institution. If employed in a 360-degree fashion, in which employees also review employers, it can offer chances for meaningful dialogue and systemic issues to be addressed on an ongoing basis. It is, inherently, not a problematic idea.

However, the way in which that performance management is carried out often results in greater issues. We can likely accept that it is important to measure performance. The more difficult – and important – question is – **how**?

To illustrate why the **how** is such a complicating factor, let's briefly discuss three widely accepted managerial truisms:

1. We value what we measure, because we often can't measure what we value.
2. When a measure of quality becomes a target, it ceases to measure quality.
3. The standard you walk by is the standard you accept.

There are lots of things that may be valued in an employee. Adherence to corporate culture is a common one – almost every institution will have a vaguely cult-like view of 'who we are as a company'. We may value collegialism, or courtesy amongst the workforce. We might feel collaboration is important or we may think competition is the best. We might value 'service' on committees or as an ambassador within the wider culture. Universities increasingly prize 'impact', which is to say evidence that the work being done within a university has a benefit for the people outside academia. Perhaps a company may praise itself based on how well it supports a diverse workforce, or that it is especially friendly to families.

The extent to which an employee contributes to this 'softer' sense of what an organisation wants to be is something that is often difficult to articulate, and especially difficult to quantify. In the absence of being able to truly measure the things we think are important, what tends to happen is the opposite. The things we **can** measure take on a sense of importance simply as a result of being amenable to metrification. That means a lot of us are assigned a lot of numbers and 'observation notes' about things that are only a distant proxy for the actual trait to which an organisation may ascribe value.

Productivity seems like it would be the easy one to capture, because surely it's straightforward to say how much of something a person has done, and how well they did it? Well, let's look back to the idea of metrics and how that would be applied to the common activities in a university – teaching and research.

How does one measure the quality of teaching? Let's simplify the problem by thinking of this as teaching in the traditional sense – a student takes a course, within a wider programme of study. That course has several weeks of instruction. At the end, the student is assigned a grade (we'll come back to that). A survey may be sent out to those taking the course, asking them to rate it according to several pedagogical principles, and then an evaluation committee may be convened to discuss those results along with student representatives from the course.

Seems robust, right? That, of course, is the illusion. We can measure all of this, so we value all of this. Or at least, we're supposed to.

Let's begin with the survey part – the specific questions that form part of this are reflective of the biases of those that construct the instrument of evaluation. Is it the job of a university lecturer to motivate students, or should a university be able to assume students are motivated by their own desire to learn? Does the survey ask if the lecturer motivated the class? How do we define motivation?

Normally what happens is that the desire to track something seen as valuable is condensed into a Likert scale, to which students assign a number. The ambiguities around what is meant are flattened down. Four out of five – that's how motivational the class was.

The same process holds true for all parts of the evaluation. Should a lecturer be part entertainer? Did they use humour to illustrate their arguments? Did they include memes? Did they conduct assessments by social media? How 'with it' are they? How many points do you deduct from 'social relevance' for a lecturer who uses the term 'with it'?

Every single question in a survey flattens nuance, encodes biases, and renders complex judgements down into a single number. Perhaps those numbers are then averaged out, perhaps the 'top line summary' of a survey is based on a simplified 'Did you enjoy the course' question. In the end, all these averaged approximations from all these students are collated into a sense of how good the course was as judged by the metrics we could gather. Do they capture value? No. Do we value them anyway? Yes.

We are not arguing here that student evaluation of courses is useless, because it isn't. This is information – in the sense we have discussed earlier in this book – that we can refine into knowledge, and over time into wisdom. However, that only works when the figures are treated as such and not when they become part of formal performance management. At the university the authors are currently employed, a course that falls below a three (out of five) on the overall satisfaction grade becomes subject to a formal process of re-evaluation and re-design. What if the response rate was 20% and the 80% that didn't answer loved the course? Doesn't matter. The metric is what drives the policy.

You probably see where we're going with this …

So, what is a professional incentivised to do here when these numbers measure only abstractions of quality, and the summation of those numbers activates formal performance measurement? Juice the numbers, of course. Okay, 75% of respondents said that the workload was too high. You can argue why the workload is appropriate and commensurate with what students will face in the real world (as in, you can stick to your guns that there is a pedagogic benefit to be had from the workload) … or you can soften the requirements of the course. Your number goes up (objectively, the course gets 'better') but the teaching value – the real, genuine, honest-to-god impact of the teaching – is worsened. Sixty percent of students say that they didn't have enough prior knowledge to follow the course 'advanced teaching of a thing'. You could raise entrance standards which students must achieve before they can take the course (and thus lower the number of students, which may result in the course falling below a threshold for cancellation; there are lots of metrics at play). You could take away the advanced content in favour of more introductory content (perhaps resulting in an overall blend of material that no longer counts as 'advanced'). Or perhaps you tell students 'I expect you to make up the difference between your starting knowledge and the requisite knowledge. I'll give you a reading list'. The latter – unquestionably more effective in giving students opportunities to ground themselves in the topic. You better believe though that making the content easier makes a 'better' course for the metrics.

That is, of course, unless you're a woman. If you are, you'll find you also need to navigate a hostile evaluation regime that punishes you for behaviour regarded as laudatory in men, and for any mistakes you make to be punished much more harshly in review (Kreitzer & Sweet-Cushman, 2022). A man can come into a lecture, hungover, and ramble on for a few hours from a vague memory of what the slides (that he forgot to bring) were supposed to be. 'Lol, what a legend!', 'Hilarious!', 'Eccentric and fun', 'really knows his stuff'. The same performance from a woman is more likely to be regarded as 'unprepared', 'slovenly', and demonstrate 'lack of structure in the presentation' (Mitchell & Martin, 2018).

Perhaps then we should look at all of this from another perspective – you could assess how many of your teachers have formal teaching qualifications, and try to raise that number. After all, if they have formal training in teaching, then they must be effective teachers? This is a thing for which students often express a desire, but the evidence shows little correlation between having a qualification and teaching well. Of course, we've already discussed the difficulties in determining what good teaching is so maybe we shouldn't put too much faith in any of this.

Really, the simplest way to improve reviews of teaching is to give out higher grades (Berezvai et al., 2021; Krautmann & Sander, 1999; Stroebe, 2020). After all, the things we've just talked about in terms of metrics apply to that ultimate summative form of evaluation of work. Teachers tend to stress to students that the feedback around a grade is what matters, but that's not the thing that gets reflected in transcripts, in degree evaluations, and in entry to other courses in a system. A grade is the ultimate flattening of nuance, and it results in students working to the assignment rather than the topic to the detriment of the wider knowledge they are supposed to be seeking. A grade, in other words, is the perfect example of valuing what we can measure.

That's why when a measure of quality becomes a target, it ceases to measure quality. It begins to inspire people to game it – we lose the link between what we were hoping to capture and instead focus on improving the numbers that are its vague proxy. The result is often to the detriment of the organisation. As always, this isn't a problem linked purely to academia. We've seen plenty examples of bug bounties that drove developers to conspire with testers to introduce more bugs into a system. We've seen 'lines of code per day' targets that resulted in software bloat and reduced performance. Social productivity measures based on 'hours worked' are at the core of many of Japan's overworking problems. Rewards based on delivering functionality in an application can lead to over-design and confusion of architecture. Tracking 'commits' can lead to code churn. Comparing velocity of development between teams results in dangerous over- and under-estimates. Financial targets can result in betrayals in the workplace as well as a downtick in performance once a target is met ahead of schedule. Paper writing targets encourage academic work to be salami sliced, as we have discussed. The world is full of gameable metrics, because as soon as they are made important then people will work out the lowest cost way of meeting them.

Finally, all of this still means nothing if standards aren't actually enforced or are conveniently discounted in favour of other accomplishments. An institution may stress the importance of collegiality and respect, but if it is constantly rewarding its most toxic 'superstar' high performer, then that is what an organisation truly values. 'The standard you walk past is the standard you accept', as David Lindsey Morrison has argued (Calil, 2021). If management turns a blind eye to dysfunction so as to protect well-regarded abusers, then the obvious message is 'be well-regarded'. That particular situation is one that is repeated again and again. The toxic work culture at ZA/UM, sexual harassment and abuse at Riot Games, Sony, Activision Blizzard, and many more. The concept of the 10xer in tech culture (a person who performs exponentially better than their baseline colleagues) has led to slew of toxicity across Silicon Valley. The 'genius' of a highly capable developer can result in management bending over backwards and overlooking complaints to keep them happy. These are the standards that are accepted from valuable workers (talent, management, and executives) because the rest of the company walks past.

'It was an open secret', McAlpine and Duncan are quoted as saying in relation to the culture at the University of Dunglen. Sometimes the problem isn't a few bad apples. Sometimes it's a whole rotten barrel that turns all the apples bad (Zimbardo, 2011).

NOTE

1 https://www.timeshighereducation.com/news/student-complaints-england-and-wales-reach-record-levels-again

REFERENCES

Berezvai, Z., Lukats, G. D., & Molontay, R. (2021). Can professors buy better evaluation with lenient grading? The effect of grade inflation on student evaluation of teaching. *Assessment & Evaluation in Higher Education*, *46*(5), 793–808.

Calil, T. (2021, September 3). Applied ethics in the military. How to mitigate ethical lapses? https://ssrn.com/abstract=4024943 or http://dx.doi.org/10.2139/ssrn.4024943

Fritsch, M., & Krabel, S. (2012). Ready to leave the ivory tower? Academic scientists' appeal to work in the private sector. *The Journal of Technology Transfer*, *37*(3), 271–296.

Heron, M. J., & Belford, P. (2016). Musings on misconduct: A practitioner reflection on the ethical investigation of plagiarism within programming modules. *ACM SIGCAS Computers and Society*, *45*(3), 438–444.

Krautmann, A. C., & Sander, W. (1999). Grades and student evaluations of teachers. *Economics of Education Review*, *18*(1), 59–63.

Kreitzer, R. J., & Sweet-Cushman, J. (2022). Evaluating student evaluations of teaching: A review of measurement and equity bias in sets and recommendations for ethical reform. *Journal of Academic Ethics*, 20,73–84.

Kwak, Y. H., & Stoddard, J. (2004). Project risk management: Lessons learned from software development environment. *Technovation*, *24*(11), 915–920.

Lopp, M., & Lopp, M. (2021). Titles are toxic: Titles place an unfortunate absolute professional value on individuals. In *Managing humans: More biting and humorous tales of a software engineering manager* (pp. 131–135). Apress.

Mitchell, K. M., & Martin, J. (2018). Gender bias in student evaluations. *PS: Political Science & Politics*, *51*(3), 648–652.

Niles, M. T., Schimanski, L. A., McKiernan, E. C., & Alperin, J. P. (2020). Why we publish where we do: Faculty publishing values and their relationship to review, promotion and tenure expectations. *PLoS One*, *15*(3), e0228914.

Ronson, J. (2016). *So you've been publicly shamed*. Riverhead Books.

Shelton, N., Laoire, C. N., Fielding, S., Harvey, D. C., Pelling, M., & Duke-Williams, O. (2001). Working at the coalface: Contract staff, academic initiation and the RAE. *Area*, *33*(4), 434–439.

Stroebe, W. (2020). Student evaluations of teaching encourages poor teaching and contributes to grade inflation: A theoretical and empirical analysis. *Basic and Applied Social Psychology*, *42*(4), 276–294.

Van de Poll, J., & Kroese, T. (2022). A new approach in extending the vitality curve to teams. *International Journal of Business Management and Economic Research (IJBMER)*, *13*(1), 2005–2012.

Zimbardo, P. (2011). *The lucifer effect: How good people turn evil*. Random House.

CHAPTER 9

Senior University Members Implicated in Growing Scandal

NEWSPAPER ARTICLE

An exclusive report by Jack McKracken for the Dunglen Chronicle

Events at the University of Dunglen took a surprising turn today when an anonymous hacker took to Blether and other social media platforms to leak a series of damaging documents from senior academics. The zip file contains numerous emails, agenda items, and confidential minutes. Hours after it had been released, the information had moved into torrents and onto the dark web, instantly ensuring it could not be suppressed by any effective means. Several of these leaked items clearly suggest that the principal Professor Sir David Tumblewood was complicit in the academic fraud that has shaken the University's reputation to its core. Other senior academics implicated in the leak include Professor Ian McManus, chair of the research ethics board, and Sir Gideon Lazenby, the university treasurer.

One of the damning emails contained within the archive is reproduced below:

```
To: John.Blackbriar@dunglen.ac.uk
From: David.Tumblewood@dunglen.ac.uk
CC: Ian.McManus@dunglen.ac.uk
```

DOI: 10.1201/9781003426172-9

Subject: re: ScotOil contracts

On Monday, Jan 20, 2020 at 3:15PM John Blackbriar wrote:

> which doesn't meet the necessary requirements for statistical significance.
> I know that you've been very keen on us making inroads with ScotOil, but I don't think it's a great idea to sell them on an algorithm which won't have anything close to the efficacy they are being led to believe.

John

I understand your worries here but I think you are being unnecessarily cautious in your analysis of the model.

Statistical significance at 0.05 is great, but it's not a binary thing. Things don't stop being interesting at 0.05p. I think 0.07 is more than enough in the circumstances to suggest that the algorithm works as we have advertised.

I've CC'd Ian as our resident 'Expert on Ethics', and he may take a different view on this, but my own experience with these kind of data points in the industry tells me that you're getting worried about what are unreliable, unrepresentative samples. The equipment used for this kind of thing operates at huge pressures deep in the ocean, and they don't always return accurate results. The preliminary data points that we gained during the pilot study look more 'real' to me.

Eliminate the outliers, fold the preliminaries into your data sets, and run the model again.I'm sure you'll see there's nothing to worry about your P values.

I don't need to remind you just how important this collaboration is for the university, and for your own research agenda.

Don't spook the horses, John.

Regards, David.

Before becoming the principal at the University of Dunglen, Sir David was CEO of the Nexus Energy Group, an oil company with extensive interest in the North Sea region.

Professor McManus' reply was included in the archive:

To: John.Blackbriar@dunglen.ac.uk
From: Ian.McManus@dunglen.ac.uk
CC: David.Tumblewood@dunglen.ac.uk

Subject: re: ScotOil contracts

On Monday, Jan 20, 2020 at 3:15PM John Blackbriar wrote:

< I've CC'd Ian as our resident 'Expert on Ethics', and he may take a different view I might, but I don't! I think you've done all you can to sanitise your data sets, John – remember, garbage in, garbage out. As long as you indicate in all publications and correspondence that you're working with processed rather than raw data, I don't see an issue with what David is recommending. Call it 'representative historical data samples'. >

Ian

A later email from Sir Gideon to Professor Blackbriar focused on the financial implications for the university:

To: John.Blackbriar@dunglen.ac.uk
From: Gideon.Lazenby@dunglen.ac.uk
Subject: re: Departmental finances

On Wednesday, Jan 22, 2020 at 2:11PM John Blackbriar wrote:

< as important as the ScotOil contract is to our bottom line. You know yourself how important this is. The consultancy fees that ScotOil are offering are enough to fund your department for five years, even taking into account the university top-slice. It's more than that though – this could be the start of an incredibly important revenue stream for the entire institution. If the initial explorations with ScotOil work out, the rest of the companies operating in the North Sea will be knocking at our door. And don't get me started on the international opportunities! See you on Thursday, >

Sir Gideon.

A close friend of Sir Gideon told the Chronicle today. 'Gideon was aware that Blackbriar was having a crisis of conscience, but he felt that John's gut feeling should override the disappointing initial results. He had a discussion in a corridor with the professor in which he essentially said "I know you're having a crisis of ethics, and I don't feel comfortable putting this in an email, but you have to trust your gut even if the statistics aren't panning out. Clean out your input data and try again"'.

These emails have come at an awkward time for the university which has struggled to escape the cloud of suspicion that has settled on the institution. Professor Blackbriar remains unavailable for comment, but Karan Chandra who is handling the defence for two students suspended under suspicion of academic fraud, released the following statement:

```
Today's revelations from the mysterious hacker known only as "Nemesis" have provided further evidence, if such was needed, of the culture of corruption at the University of Dunglen. They show senior academics conspiring to commit academic fraud to ensure that revenue streams would not be compromised as a consequence of unpromising results. It's a shoddy deception from start to finish.
```

An analysis of the papers published by Blackbriar and his colleagues has shown that he did indeed refer to his analysis as being performed on 'representative data samples' as directed, as can be seen in this extract from the paper 'An Aggregated Analysis of Algorithmic Applicability':

```
Application of the algorithm to a subset of representative historical data samples demonstrates a strong correlation between recorded fitness of generated solutions and later validated extractable reserves. A statistical analysis using a Student's t-test shows significant results t(1008) = 5.212, P < 0.05 for all aggregated sets.
```

However, nowhere in any of the papers or supporting literature is the mechanism by which data points are selected for inclusion published.

'Reproducibility is an important element in science', research ethics expert Professor Callum Sunderland at the University of Alba tells us, 'And part of that is allowing external parties to have access to the data sets you used and the criteria by which data points were selected. Commercial interests in this particular instance meant that the first criteria of this could not be met, and the second criteria seems to have been wilfully not met. It's not straight up fraud, but highly suggestive of a desire to avoid detailed scrutiny'.

When asked about the inclusion of pilot data in the model, Sunderland adds, 'I can't imagine a situation where that would be appropriate. Pilot data is by its very nature unlikely to have the methodological rigor that would be needed to combine it with the operational data gathered. Sir David may be correct in his critique of the data, but there's no way to tell if his gut feeling on the matter is reliable. It's not a good idea to base a research conclusion on what "feels right"'.

He concludes, 'A lot of bad science is out there because people followed what they thought the results should be rather than believe what their own analyses told them'.

The zip file from which these emails were obtained is currently freely available as a torrent on the Internet. It contains several sub-archives, each of which is password protected. The torrent is currently seeded by several hundred people, suggesting that thousands of copies of the archive may be out in the wider internet. The mysterious hacker at work behind the pseudonym Nemesis has also released a statement of sorts – a message that was scrawled across the University's web page for several hours this morning.

'This is only the start. A new password will be revealed every three days until justice is served'.

We, and the senior management of the university, await further developments at the institution with interest and apprehension.

PUBLISH AND PERISH

The fish rots from the head down, as someone probably said. Our newest article on the senior university members at Dunglen shows the root of some of the systemic issues we have been exploring.

The phrase 'publish or perish' codifies the expectations placed upon academic scholars as effectively as anything for which we might hope. Most within academia have encountered the phrase, and more still in research-intensive institutions have felt its impact. It expresses the common expectation that institutions have of their researchers – publish work, often. It is not enough though to simply publish, it is necessary to publish work in venues that will lead to citation, to name recognition, and to the often ephemerally defined impact (Donovan, 2011) The pressures associated with this have increased in modern years. Nothing happens in a vacuum. Competition over academic jobs has become ever more fraught because universities continue to produce more qualified academics than the system can consume (Powell, 2015). This means that the expectations on those seeking permanent faculty employment end up being continually revised upwards. Uncertainty and instability in employment creates a perversely effective incentive system on those scholars least able to challenge it (Shen, 2015) and this in turn creates a powerful 'free market effect' in the academy. If the academy wants high-performing academics, it can select for that through evolutionary pressures.

On the surface, leaving aside the human impact of often unreasonable expectations, this might even seem like a good thing. The result after all is a perpetual raising of the bar of scholarship so that the next generation is notably better than the last. Evidence suggests though that this has not quite worked in the way that might be hoped. When quantity is incentivised, then other things will suffer as a result. When a research infrastructure emphasises citation, it will produce a culture of citation that is independent of the merit of the work. Previous chapters have discussed some of the ways in which citation can be manipulated to create the external appearance of impact without any of the concomitant benefit for society. Instead of publishing one truly landmark paper, young academics will often aim for the 'Minimally Publishable Unit' (Budd & Stewart, 2015) – the smallest measurable amount of information that would pass the criteria for publication in an appropriate conference or journal. Four smaller papers can often appear more valuable on an academic resume than one large paper and almost always serve to skew a citation count upwards as a simple consequence of cross-referencing. The Scandal in Academia itself is an example of that. It is collected here as a unified monograph of over a hundred thousand words, but it was initially presented as a series of papers each generating its own ecosystem of citation. We have already discussed how easily this can be abused, and then went on to abuse it in exactly that way.

Work may be 'salami sliced' so as to appear several times in different publication venues – perhaps with a change in emphasis or a change in advertised relevance. It's hard to quantify the extent to which this happens though because the nature of different academic fields prevents meaningful analysis. The number of papers considered 'enough' varies wildly from field to field and university to university. While this all may seem harmless, it has an impact on the replicability and reliability of scientific results; statistical significance can be skewed if one single study contributes multiple data points via several articles.

However, the pressure to publish has a more corrosive impact on the quality of the work. Results are often over-inflated or over-stressed. Statistical analyses can be 'massaged' to give a value for significance that is appropriate for publication – sometimes known as p-hacking (Head et al., 2015). This can also occur when it comes to submitting articles of peripheral relevance to a publication. The text can be manipulated, or 'contextualized', to make the work appear more directly suited to the scope of the venue that it can legitimately claim. Negative results may be recast as positive findings. None of this is academic fraud of the kind discussed previously – merely a small case of 'refining' what was discovered until someone will pay attention.

Consider for example a set of data that yields no statistical significance when viewed through a standard analytical test – a T-Test or a Chi-Squared test for example. By applying ever more sophisticated tools to the same data, it becomes increasingly likely that one will reveal a statistically significant result, purely because of the way in which chance manifests in data analysis. A cynical peer reviewer might ask 'why didn't you get significance from a less obscure testing regime', but it might also be interpreted as a sign of finding something of genuine worth that was hidden beneath the surface. It's not that the reporting is wrong or even misleading. It's just that if there is a pressure to publish people can be relied upon to find publishable data if they look with sufficient diligence. We often consider statistical significance with a P value of 0.05 to set a benchmark for 'minimal significance' for publication (Nuzzo, 2014). As David Tumblewood says in the hacked email chain, 'things don't stop being interesting at 0.05p'. The nature of significance is often over-stated, and it's entirely possible that a significance of 0.07 is substantive enough upon which to build a case for relevance. After all, the 'gold standard' of significance is that there is only a one in twenty chance the results were derived through random chance. A significance value of 0.07 suggests only one in 14.28 odds that the results were derived through chance. Who wouldn't back a horse at those odds, especially if you could look at the horse again from a different angle and speed it up accordingly?

Sometimes the data is massaged in different ways – we've discussed the extent to which outliers can be easily discounted from an analysis to bring the figures in line with the baseline for publication. This is a common, and indeed necessary, part of data curation. However, the fact of it matters less than the motivation – good science should be impersonal and not incentivised to find a particular result. With the Blackbriar Algorithm, we're dealing with data points derived at the extremes of what mechanical and digital equipment can gather. It's not unreasonable to remove obvious outliers, but the extent to which they are 'obvious' can be influenced by the external pressures to end up with publishable data. Negative results are incredibly important in science, but they are rarely published in certain fields (Matosin et al., 2014) and so researchers are disincentivised away from authenticity in their analyses. If you keep looking you might find something positive, so why not keep looking until you do? The negative result may be more authentically valuable but the positive result is more **publishable** even if it may end up being misleading.

These are largely internal pressures that come from the publish or perish mindset – an urge to find the publishable units in the work you do because otherwise you're not going to keep up with a tempo of publication set by those competing for the jobs you want. There are external pressures too, often set at an institutional level. Many academics work within fixed expectations of publishing – either formal parts of their contract of employment or an informal expression of their local culture. 'Two papers per year in high impact journals' is a common benchmark in research-focused UK universities, even if it's rarely formally expressed. As it is with many norms in a highly mobile workforce like academia, it's sometimes more a diffusing of hearsay than a formal contractual obligation. We carry the expectations of our former workplaces into every new one we enter.

On top of this we can layer formal expectations that influence expectations of publication – papers are what get you academic credibility, and credibility is what finds you collaborators. Collaborators are what you can use to get grant funding, and grant funding is, eventually, how you pay for your salary. Research active academics cannot simply focus on papers, but those papers have a role to play in a funding landscape that is intensely competitive. It is cynical to say that universities are only interested in research funding, but it's hard to deny that's how they often behave. More publications equal more credibility which equals more funding. The value of the scholarship often seems to have diminished value in comparison to the mere fact of its publication.

Consider for example the subtle, and not-so-subtle, hints that Professor Blackbriar receives in the hacked correspondence. 'I don't need to remind you just how important this collaboration is for the university, and for your own research agenda', 'The consultancy fees that ScotOil are offering are enough to fund your department for five years', 'Don't spook the horses'. It's not the case that anything as simple as managerial pressure is being brought to bear here – it comes in a more insidious, more plausibly deniable form. It comes as well-meaning guidance that is looking out for Blackbriar's best interests if he could just shake off his understandable uncertainty about the results. 'As long as you indicate in all publications and correspondence that you're working with processed rather than raw data, I don't see an issue with what David is suggesting', says the chair of the research ethics board. He's right too – there's no academic fraud if you're honest about everything you're doing. However, there's a cynical degree of 'knowing' in advice like this – delving into data sets for replicability happens infrequently and given the commercially sensitive nature of the data it's not the case that even those directly working on the project could assess the validity of the processing. We see that too in the Blackbriar emails leaked by Nemesis. When most journalists and managers will read the top-line only, they rarely have the time or expertise to consider the implications of a disclaimer buried in the text.

It is this more than anything else that sets the behaviour of Blackbriar and his colleagues in context. We've discussed the cases of Diederik Stapel (Markowitz & Hancock, 2014), Yoshitaki Fuji (Normile, 2012), and Dipak Das (Retraction Watch, 2012) as outright cases of fraud and academic misconduct. Here, everyone is telling the truth. They're just not telling the whole truth. They have found a horse and said it was a zebra because if you looked at it at a certain angle, in the right light, you could 'kind of' see some stripes. This can often happen through over-enthusiasm on the part of an amateur – indeed the word 'Zebra' is a kind of medical slang in itself. 'When you hear hoofbeats, think horses and not zebras'. As in – when you observe a set of symptoms, assume the most common underlying disease. Sometimes invoking the possibility of a zebra though is more conscious and intentionally manipulative.

It's also important to note here that part of the guidance received by Blackbriar was an exhortation to 'trust your gut'. Widely understood, but rarely explicitly spoken, is the expectation that research is to some extent guided by intuition. We cannot discount here that Blackbriar is a professor of considerable renown and experience in his subject area. While in the end we must always yield to evidence, there is a grey area where we are still trying to intuit what the evidence is actually saying. In this case, it can be argued that Blackbriar is being pressured into making claims that the evidence cannot reasonably bear – but it could also be argued that this is a necessary part of divining whether this evidence is actually answering the question we think we asked. Perhaps the problem isn't with the analysis – perhaps it genuinely does lie with the data collection. Much of the conversation we see is **suggestive** of a conscious attempt to mislead. However, it can also quite easily be interpreted as a professional, appropriate discussion on the nature of evidence. Sometimes we may see flaws in results that we cannot necessarily articulate. Experience is a powerful guide to where we may see meaning in aberrations, but it can also be something that calcifies our thinking and makes us resistant to change. Part of the job of an academic is to know the limits of their own epistemology. The 'plea to authority' is often considered a logical fallacy, but to dismiss expertise as irrelevant in interpretation is to succumb to the same fallacy in reverse. Sir David Tumblewood too has years of experience as an oil man working with data in these kinds of extreme contexts. It would be unwise in the extreme to discount his invaluable opinions wholesale. On the other hand, as noted philosopher Space Ghost once said, 'I believe every word that man just said, because it's exactly what I wanted to hear'.[1]

BLOWING THE WHISTLE, AND WHY

One thing that we haven't yet discussed in this book is the expectation of disclosure implied in that cited email chain. We now live in a world of largely frictionless data transfer – the increasing capacity and speed of modern computers has been more than a difference in scale, it's a difference in kind. Fifty years ago, it would not be possible for Edward Snowden or Chelsea Manning to download gigabytes of sensitive information and make those collections easily available to journalists and activists (Heron, 2016) – the act of removing that amount of data from a secured facility would have involved physical media and a supply of wheelbarrows. Now, it's a matter of a few minutes to transmit vast amounts of information across the globe. This has had a remarkable impact on public discourse and the expectations we have of transparency for our governments. 'Information wants to be free' is a common slogan of free speech and free information advocates, and when information becomes free, it is very difficult to chain it. Consider for example the encrypted torrent file Wikileaks distributes as insurance against censorship (Cammaerts, 2013) – in the event anything happens to their central repositories, all they need to do is release a password and everything contained within those distributed archives will become public knowledge.

Along with the difficulty in containing information once it is revealed comes increased difficulties of preventing its revelation in the first place. Academics and public figures are often the targets of malicious, or merely inquisitive, attempts to access their private data. Especially that which is primarily stored on cloud services or on remotely accessible servers. Unintended disclosure is not a new problem – journalists of the less reputable outlets have been known to trawl through the garbage of celebrities in the hope of finding reportable information. The whole business of espionage is conducted in an ongoing exercise of escalating revelation – using easily accessed information to lever open access to other more restricted pieces of information. However, the ease at which this can now be done remotely, and the quantity of data that can be obtained, has changed the texture of privacy online.

Consider for example the consequences of a badly chosen and easily guessed password. If used too liberally between multiple services, a malign agent might use it to unlock access to years of email, social media accounts, chat logs, photo repositories, and more. Many of the things that might be revealed could be incriminatory or open further access to external parties involved in discussions.

We all use many digital services on a daily basis, and few of us know for sure how diligently these services are guarding our data. The best services are resistant to hacks and leaks, storing sensitive content in an encrypted format and ensuring it does not yield itself to brute-force password attacks. With the growth of encryption came a corresponding increase in the sophistication of hacking. As an example, to counteract the time taken to compute password hashes pre-computed rainbow tables became a common tool of those looking to force access where they were not invited. That in turn led to the practice of 'salting' a database so that rainbow tables became much less effective. The war between those that want access to data and those that wish to prevent that access is an escalating one, with new fronts opening on a daily basis.

However, institutions and services cannot necessarily afford to be in the vanguard of the war – it is costly, requires expertise, and is often highly contextual and situational. As a result, we are all more vulnerable to unwanted disclosure than we might like. A hack at AdultFriendFinder in 2014 revealed the sexual proclivities of over 400m users. In 2013 and 2014, Yahoo was responsible for leaks that impacted one and a half billion users in total. In 2015, hackers stole gigabytes of user data from the site Ashley Madison, which sold itself as a venue for married adults to have a discreet affair. Equifax was responsible in 2017 for a massive leak of incredibly sensitive financial data – around 150m records. In 2021, hackers managed to gain access to Microsoft's Exchange servers and reveal commercial and governmental emails. Also in 2021, Facebook experienced a data breach that resulted in exposing the names, phone numbers, accounts, and passwords for over half a billion users. In 2023, classified war plans relating to the US, NATO, and Ukraine were revealed on social media. The true impact and severity of these leaks varies from case to case, but it's clear that we cannot entirely trust other parties to look after our data for us.

The impact that these kinds of disclosures has can be seen in the articles we've been referencing through this chapter. It is often not the case that what we have put online is sensitive information in and of itself, but rather information that has a certain expectation as to the context of its audience. Within a technical publication, we might reference jargon, common tools, and techniques and make reference to technology and concepts that could not be expected of a 'lay' reader. That's entirely appropriate because academic papers are intended to be read by a specialist audience, in the same way that procedural motions in a court are intended for a legally literate audience rather than the populace as a whole. We would not use such specific terminology when describing the work to friends and relatives with whom we cannot assume common ground in understanding. There is a context of audience that we all implicitly function within. Or rather, there are multiple complex contexts that overlap in ways predicted by technical experience and demographic makeup. Clashes of these contexts can yield misunderstandings and misattribution of intention.

Even in more informal communications, there is a form of social context that textures and informs the way in which we discuss topics. This context is rarely explicitly referenced when leaked emails and such are published for the unguarded and salacious gossip they contain. Communication is not simply statements delivered in order – it's a collaborative exercise in building understanding. Both the source and recipient of information are engaged in a collective task of interpreting meaning (McCulloch, 2020), and the literal definition of words and terms by themselves may not fully illuminate that intended meaning.

Consider for example the simplest of these contexts – familiarity. In-jokes and shared experiences can radically change what words mean and how they should be interpreted. A shared bond between people can ensure that jokes are understood as such, and sarcasm is picked up for what it is. In taking a random email from a random inbox, we are ignorant of such nuance and it's impossible to codify it in a form that genuinely captures the often heavy weight of contextual subtlety. An email that states 'you could always fake the data' might be intended as a joke when sent to the designated recipient. The recipient might understand that instantly because of social cues not encoded into the missive. No serious intent or suggestion of fraud is intended. However, when read uncharitably by an otherwise dispassionate independent observer, it looks an awful lot like an exhortation to misconduct. More than this, attempts to explain the context almost always come across as insincere justifications – when our statements are distributed wider than our intended audience, we are put in an unwinnable situation because the attempt to justify such content – which is being assessed in stark isolation from its intended usage – will almost always seem defensive. Consider for example Professor Barry Spurr (Davis et al., 2014) who resigned from his position as a professor at the University of Sydney after a leak of racist and sexist emails. His defence that the emails were a 'whimsical word game with colleagues and friends' convinced few.

Whether that is the case with the Blackbriar emails, we will never know. Is the hectoring tone of publication pressure a jovial reference to their shared contempt for the philosophy of publish or perish? Perhaps ScotOil is something of a joke between the participants in the conversation. In that case, 'I don't need to remind you just how important this collaboration is for the university' might be a caustically arch statement that carries the opposite meaning of what the words say. The use of the phrase 'representative historical data samples' is in quotation marks. Is that an in-joke? Does it refer to a draft institutional policy? We don't know and can't know, and as such we should be wary of reading too much intent into such emails. It's common for the misquoted to claim 'out of context' for statements they are forced to defend in the public sphere – that's not true in this case, because the full surrounding context of the remarks is available. Or at least, it is potentially available if someone wants to download the torrents referenced. However, context can incorporate more than the words that precede and follow the indicated statements.

With our mysterious hacker too, we must appreciate that there may be a hidden agenda at work. The nature of an informational leak is that it is asymmetric – we cannot know how much information has been leaked, or what impact the order in which it is released is intended to have. The very nature of revealing hidden information is shrouded in secrecy because we do not know what remains hidden at the end of the process. We do not know if this is the full set of information, or whether later emails would contradict or perhaps further contextualise the earlier ones. Could we honestly say, in all fairness, that this leak even related to the project in question?

We are led to believe that it is the Blackbriar Algorithm that is being discussed by the dates and the juxtaposition of the emails with the ongoing coverage. Perhaps this is Blackbriar commenting on the work of another colleague. We simply don't know because we have been given cherry-picked emails that are devoid of their full, unambiguous situation within a conversational record. They certainly seem suggestive, but that does not constitute evidence in its proper sense. Consider for example the Climategate (Leiserowitz et al., 2013) emails. These had been obtained through the hacking of a server at the Climatic Research Unit at the University of East Anglia. They were heavily publicised in the climate change denialist press as evidence of a sustained conspiracy with regard to the integrity of climate change research.

And to be fair certain emails could be interpreted in such a way as to suggest conflation of conjecture and conclusion, manipulation of results, and scientific trickery of the kind outlined in the Blackbriar case we have been discussing. Eight independent local and national committees investigated the emails and found no evidence of fraud or scientific misconduct. The informality of communication and the coining of internally consistent jargon permitted sceptics to read more into the communication than a fairer, more realistic reading would permit. The Guardian, reporting on the scandal, noted that there was much to suggest that the emails released had been filtered to give undue prominence to a handful of emails that could be most easily interpreted in an unfavourable light. Terms such as 'Mike's Nature trick' were used to describe methodological techniques that could meaningfully connect together two different kinds of data sets (Skrydstrup, 2013). Within the expected audience of the emails, the term was understood to refer to a broadly understood and accepted analytical practice. Outside of that context, it took on a much more sinister tone.

This shines a light on why whistleblowing is not necessarily an ethically pure phenomenon, especially when it is being done by an anonymous party. There are many reasons why disclosure of identity would not be desirable when reporting on wrongdoing. It would be churlish to assume all those that maintain their distance from disclosure are doing so for dishonest reasons. However, the fact that the action is reasonable does not inoculate it from its troublesome aspects. An anonymous party may have no agenda, or they may have a very pointed agenda. They may have revealed all the information, some of the information, or some highly targeted subset of the information. They may have released the information unedited and uncurated, or they may have edited key elements of the data to further their own agenda. They may have unfairly focused attention on one participant in a conversation while shielding others. Within the Scandal in Academia, we do not know who Nemesis is – at least at this point. The agenda that they harbour is something we need to know before we can assess the reliability of the information in its fullest light.

Data security is especially important within professional spheres because we are often operating under formal requirements for data management. We may be beholden to funding partners to protect commercial interests. We may be obligated by a research policy to make available full data sets in their unedited and uncurated form. We may have an ethical duty of care to ensure the anonymity of research subjects and participants, and we may have a similar ethical duty to ensure the protection of their contributions. We will also almost certainly be under a legal responsibility to handle data in line with existing legislation – the General Data Protection Regulation (GDPR) Data Protection Act and the Freedom of Information Act in the UK being obvious examples.

And yet, we are often pulled in multiple directions – we need to lock down information for protection while opening it up for collaboration. Organisational network architectures are often insufficient for dealing with the flexibility and variety of data needs and professions often resort to the use of external tools – drop boxes, torrents, servers outside of institutional firewalls, GitHub repositories, and more. This is often as much for necessity as it is for convenience, but every service that we use adds another point of weakness for those looking to gain access to information over which we would prefer to have control. Often the disclosure of the data would present no serious or significant problems – it's often hard enough to get people to pay attention to our work even when we're screaming about it to everyone we can at every opportunity. As the joke goes, 'You don't need to worry about anyone stealing your ideas. If they're any good, you'll need to force them down peoples' throats'.

Occasionally though, even the most innocuous of data points can be compromising if wielded by someone intending on doing damage and unconcerned if the damage may only be temporary. The initial attack may end up as front-page news, after all. The retraction is most likely to be a small notice buried in the back pages of a later edition. As such, what we see here in the Scandal in Academia is **perhaps** indicative of an ethical lapse from all participants in the management structure. It might just as easily be an out of context attack from an aggrieved party with their own axe to grind. Whistleblowing is a double-edged sword for everyone involved.

NOTE

1 https://knowyourmeme.com/photos/1667004-reaction-images

REFERENCES

Budd, J. M., & Stewart, K. N. (2015). Is there such a thing as "least publishable unit"? An empirical investigation. *LIBRES: Library and Information Science Research Electronic Journal, 25*(2), 78.

Cammaerts, B. (2013). Networked resistance: The case of WikiLeaks. *Journal of Computer Mediated Communication, 18*(4), 420–436.

Davis, M. W., et al. (2014). The persecution of Barry Spurr. *Quadrant, 58*(12), 20.

Donovan, C. (2011). State of the art in assessing research impact: Introduction to a special issue. *Research Evaluation, 20*(3), 175–179.

Head, M. L., Holman, L., Lanfear, R., Kahn, A. T., & Jennions, M. D. (2015). The extent and consequences of p-hacking in science. *PLoS Biology, 13*(3), e1002106.

Heron, M. J. (2016). *Ethics in computer science.* New Publisher Required.

Leiserowitz, A. A., Maibach, E. W., Roser-Renouf, C., Smith, N., & Dawson, E. (2013). Climategate, public opinion, and the loss of trust. *American Behavioral Scientist, 57*(6), 818–837.

Markowitz, D. M., & Hancock, J. T. (2014). Linguistic traces of a scientific fraud: The case of Diederik Stapel. *PLoS One, 9*(8), e105937.

Matosin, N., Frank, E., Engel, M., Lum, J. S., & Newell, K. A. (2014). Negativity towards negative results: A discussion of the disconnect between scientific worth and scientific culture. *Disease Models & Mechanisms, 7*(2), 171.

McCulloch, G. (2020). *Because internet: Understanding the new rules of language.* Penguin.

Normile, D. (2012, April 10). A new record for retractions. *Science Insider (American Association for the Advancement of Science), 2.* https://www.science.org/content/article/new-record-retractions

Nuzzo, R. (2014). Statistical errors: P values, the 'gold standard' of statistical validity, are not as reliable as many scientists assume. *Nature, 506*(7487), 150–153.

Powell, K. (2015). The future of the postdoc. *Nature, 520*(7546), 144–148.

Retraction Watch. (2012). *UConn resvatrol research Dipak Das fingered in sweeping misconduct case.*

Shen, H. (2015). Employee benefits: Plight of the postdoc. *Nature, 525*(7568), 279–281.

Skrydstrup, M. (2013). Tricked or troubled natures? How to make sense of "Climategate". *Environmental Science & Policy, 28*, 92–99.

CHAPTER 10

Drunken Professor Lashes Out in Twitter Storm

NEWSPAPER ARTICLE

An exclusive report by Jack McKracken for the Dunglen Chronicle

Professor John Blackbriar, the academic at the centre of the growing BrokenBriar Affair, ended his month-long silence last night with an unexpected and stunning barrage of social media messages. Blether is a social media platform that focuses on a form of micro-chatting – all Blether messages must fit into a 200-character limit, enforcing brevity of discussion. Messages can come with tags, which are then used by the underlying algorithm to identify trending topics – the tag employed by Blackbriar was 'DunglenStitchup', which has become a shibboleth for his growing army of supporters. In an hour-long Blether storm, he railed at senior academics in the university, his research students, and ScotOil executives all of whom he said knew exactly what was being done with the data. When queried online about his comments by bewildered followers, he confessed to being drunk and insisted he was the victim of an elaborate 'stitch-up'. He would later claim that his account has been hacked and that all these messages should be disregarded. Extracts from his tirade are reproduced below:

@DrJohnBlackbriar 8h
@DunglenUniversity pick up the dam phone

DOI: 10.1201/9781003426172-10

@PrincipalDunglenDavid you cant shut me out
DunglenStitchup
@DrJohnBlackbriar 8h
@DunglenUniversity pick up the phone pick up the phone pick up the phone
DunglenStitchup
@DrJohnBlackbriar 8h
@DunglenUniversity im not taking the fall for this, you all knew what was happening
DunglenStitchup
@DrJohnBlackbriar 8h
@DunglenUniversity I was the one who said that we couldn't publish the results, you told me in no uncertain terms I had to
DunglenStitchup
@DrJohnBlackbriar 7h
@DunglenUniversity you know those leaked emails are only the start. What about when they publish the one where you threatened me with sanctions
DunglenStitchup
@PippyBungle 7h
@DrJohnBlackbriar
@DunglenUniversity holy shit, does your lawyer know you're on Blether?
DunglenStitchup
@DrJohnBlackbriar 7h
@PippyBungle
@DunglenUniversity screw my lawyers, all they do is tell me to keep quiet while I'm slandered
LyingBastards
DunglenStitchup
@DrJohnBlackbriar 7h
@PippyBungle
@DunglenUniversity I've kept quiet for weeks! What good is it doing me???
LyingBastards
DunglenStitchup
@DrJohnBlackbriar 7h
@PippyBungle
@DunglenUniversity Maybe I should do what everyone else is doing and leak stuff to the Chronicle.
LyingBastards
DunglenStitchup
@DrJohnBlackbriar 7h
@PippyBungle
@DunglenUniversity Do you want me leaking stuff to the Chronicle
@PrincipalDunglenDavid? I don't think you do.
LyingBastards
DunglenStitchup
@PippyBungle 7h
@DrJohnBlackbriar
@DunglenUniversity are you drunk?
@DrJohnBlackbriar 6h
@DunglenUniversity
@PippyBungle Getting there!: –
D SingleMalt
DunglenStitchup
@DrJohnBlackbriar 6h bastards just blocked me but I won't stop. everyone knew, I told everyone.
DunglenStitchup
@DrJohnBlackbriar 6h mcalpine and duncan knew the review criteria for data they even helped design it.
SaveYourOwnSkins
DunglenStitchup
@DrJohnBlackbriar 6h principal senate and court knew and told me that my whole department would be axed if the scotoil stuff didn't go through
MoneyMoneyMoney

DunglenStitchup
@DrJohnBlackbriar 6h Scotoil got access to first drafts of papers and said to massage the figures to mollify shareholders
HangTogetherOrHangSeparately
DunglenStitchup
@DrJohnBlackbriar 6h and in all of this I still stood my ground and published the truth
NotYourPatsy
DunglenStitchup
@DrJohnBlackbriar 38s Sorry folks, my Blether account has been hacked. I'm going to shut it down while my lawyers and I investigate. Please disregard all messages.

Lawyers for John Blackbriar were quick to respond to the Blether storm. 'Professor Blackbriar is a subject of intense national and international attention at the moment.

The comments made are the result of mischief-making by an unknown hacker looking to implicate our client in an incident of professional indiscretion. When the individual known as Nemesis is actively leaking alleged documents from the University, it is only to be expected that their campaign would extend to parties of interest in the investigation. These comments should be ignored, as the Professor was not in control of his account at the time of the Blethers'.

Karan Chandra responded to the statements made about his clients by saying 'The allegation that my clients knew about the way that the data was being doctored is entirely false. This is merely extra evidence for the culture of bullying and buck-passing within the University of Dunglen. We will be seeking to have these comments appended to the legal complaint that we have already put in motion'.

When asked whether comments from a hacked Blether account will add any weight to that complaint, Chandra said, 'If it was hacked, how come the police haven't been approached by Blackbriar? There's every possibility, given the evidence, that he was drunk, sobered up and then panicked and invented a hacking attempt. Until the police have investigated the incident, denials like this mean nothing'. Police Caledonia confirmed to the Chronicle today that they had not yet received any correspondence from the professor's lawyers regarding the alleged hack.

Professor John Blackbriar himself has not been seen on Blether since his last message above, and the account has since been deleted. Not before several enterprising individuals captured the Blethers in screenshot form and made them available through other social networks. It is these captured screenshots that we reproduce above. Since the tag began to trend, Blether has reported a steady uptick in commentary around the scandal. Researchers at the University of Alba employed sentiment analysis – a form of natural language processing – to categorise messages posted around related tags. The results suggest that 60% of messages are supportive of the professor, 20% are negative, and the rest ambivalent. However, the analysis also identified that a large proportion of comments were from accounts with distinctly suspicious profiles – many of them created shortly after the scandal began to unfold, and that demonstrate little engagement with the platform or its community beyond this specific issue. However, it is clear that this outburst has done nothing to stem the flow of online commentary, with national influencers turning social media into a form of 24-hour news channel devoted to the topic. International commentators have also started to pay attention to what's happening at this previously obscure Scottish university.

Blackbriar himself has become a popular meme, with several meme templates around his activities entering into public discourse. Along with this, videos of his public lectures have been used as fodder for deep fakes – a form of AI-enabled video editing that transplants the face of someone from one video onto another. We have seen videos of Blackbriar playing the part of Hitler in his final moments in the movie Downfall. We've seen him take over the role of Mr Bean sitting an exam. Increasingly, it is becoming difficult to identify what is real and what is manufactured.

It is not Blackbriar alone who is the subject of such online mockery though, although in the cases of other figures in the scandal, the tone is much darker. Explicit deep fake videos of Sharon McAlpine have been uploaded to aggregators of pornographic movies, and she is also taking the brunt of negative commentary online. The University of Alba finds a 55% negative sentiment associated with Blether messages which tag her specifically.

If the Blether comments genuinely do come from the professor, they represent an extraordinary escalation in the rhetoric of the scandal. Previous emails seen and published by the Dunglen Chronicle have shown the principal and other senior members of the university implicated in the fraud. Professor Blackbriar's allegations have intensified the scrutiny being aimed at their involvement. The anonymous hacker known as Nemesis today leaked another email, which seems to back up the professor's claim about the pressure Sir David was applying to Blackbriar's department:

```
To: John.Blackbriar@
dunglen.ac.uk
From: David.Tumblewood@
dunglen.ac.uk
Subject: Financial
Pressures

Look John

We all admire your integ-
rity. We all admire your
dedication to publishing
the results of the algo-
rithm as they stand rather
than when they are cor-
rected. However, you're
doing nobody any favours.
```

Let's look at it pragmatically - much of your department's revenue comes from ScotOil's consultancy and research contracts - so much revenue in fact that I don't see how you can replace it.

I can certainly tell you it won't be replaced by central university funds. You know the score - support your centre or lose your centre.

This isn't just your job on the line. According to my figures you have six postdocs, ten PhD students and two research fellows. Are you thinking of them when you take this hard-line stance? Who exactly are you helping here?

Remember too that Alba are working on their own

AI-enabled system that will be competing directly with ours. They don't have the - what did you call it, 'tech debt?' - that we have. They can make use of all the newest and shiny techniques without inertia. Our advantage is that we are - at least at the moment - the first movers. That won't be true forever.

If you're worried about our relationship with ScotOil, all I can say is that they've got as much invested in making this project a success as we have. Their executives have already been on the phone to me saying that they need to see an 'unqualified success' from the results or they'll pull the funding. I know from deep personal experience - that's code. They want to *see* it, the *actual* results can be more nuanced.

You know what they say about lies, damned lies, and statistics.

In the end it's your call. But be aware of the ramifications if you choose to publish from a data-set we know isn't completely reliable.

David.

ScotOil have so far refused to comment officially on this email, but our anonymous insider explained: 'There are a few people who have been pushing the Blackbriar projects heavily, and we've invested millions over the past five years. ScotOil is like any company – it's full of politics and infighting and manoeuvring over status. It's certainly true that they needed a win. Whether they needed it enough to nudge the university into publishing a report that overstated the benefits of the work – well, that's debateable'.

While this strengthens the claim made against the university by Professor Blackbriar, further leaked emails seem to contradict one of the key claims made by the professor regarding Mrs Sharon McAlpine and Mr. James Duncan:

```
To: John.Blackbriar@dunglen.ac.uk
From: Sharon.Mcalpine@dunglen.ac.uk
Subject: Data Sets

Professor Blackbriar,

We're getting close to a draft of our paper for you to sign off on. Before we do that though we really do need to get access to the selection criteria for the data sets so we can properly contextualise the results.

I know you say they are commercially sensitive, but we all need to be above reproach when reporting on it. Caesar's wife must be above suspicion, and all that.

Thanks,

Sharon
```

This had the reply:

```
To: Sharon.Mcalphine@dunglen.ac.uk
From: John.Blackbriar@dunglen.ac.uk
Subject: re: Data Sets

Hi Sharon

You'll need to publish without access to the selection criteria. The criteria make no sense without access to the full data sets and would actually be misleading. I don't have permission to make those available to you for review, and even less permission to make them available for publication.

Suffice to say that the criteria have been designed to ensure maximum integrity and reliability of the data, and that they have been peer reviewed by the highest authorities in the sector and the university.

You can contextualize based on that.

Hope this helps, John.
```

The email evidence certainly suggests that the professor withheld the selection criteria from his students, which would suggest McAlpine and Duncan had no role in their construction as had been claimed on Blether.

Blackbriar's claim that he published the truth has come under criticism from several quarters. Expert in research ethics Callum Sunderland says, 'It's difficult to say really – he may have published the truth, but it was far from the unvarnished truth. Given the extent of the manipulation of the research data sets, it seems like an extensive explanatory note would have been required to fully educate readers. Cherry picking of results is not uncommon in research, and there is an inbuilt bias against publishing negative results. These things happen all the time, but that doesn't make them right'.

TANTRUMS AND TEMPTATION IN SOCIAL MEDIA

It's not uncommon in modern times for those at the centre of a media storm to lash out on social media platforms. As unwise and unhelpful as it usually is, such actions permit a degree of control over the narrative that is otherwise unavailable to an affected party. The frictionless nature of the digital age that we discussed earlier also extends to social media platforms – a single tweet can become a global discussion point in the time taken for one person to fly from New York to South Africa (Ronson, 2015). Most social media platforms are built upon a principle of sharing, and designed to make sharing as easy as possible. Facebook has its share button. Twitter has its retweets. At the press of a button a signal can be boosted by a reader, and then boosted by their readers, and so on. Much of the battle for attention online is fought in a contentious marketplace of trending topics where there are no clear strategies for ensuring who the audience will be, or how the topic will be interpreted. The 'virality' of content is a marketable trait of considerable value when wielded wisely (Hansen et al., 2011). However, the internet has a tendency to flatten nuance and dilute subtlety, and as we discussed in our previous chapter even coming to a consensus view regarding meaning is a complex act of negotiation.

With this in mind, it's common for aggrieved parties to wish to state their side of the story. Such attempts only rarely go well – often by the time a situation has gotten to this point it has already been spun into a dominant narrative by those participating in conversation. Nobody on social media gets an opportunity to 'control the story' – at best they can try to nudge it onto a more agreeable trajectory. Such attempts at adjusting the orbital dynamics of online discourse require a careful touch and more than a little bit of savvy in internet culture (McCulloch, 2020).

The temptation though can be hard to resist when one otherwise feels themselves silenced or censored by more traditional media outlets. The online discourse is dangerous ground to tread in though – it can be febrile and tends to amplify the extremes of conversation through anonymity

(Lee, 2007), deindividuation (Williams & Guerra, 2007), and simple weight of numbers. These elements come together without engineering but can also be corralled and directed as a specific unit of highly tailored harassment (Heron et al., 2014).

In 2023, it was estimated that Twitter had an active user base of around 200m daily users, with 450m being active on a monthly basis. That may have dipped somewhat as the platform rebranded itself, mystifying many, to X. With such numbers even if we consider a particular event to be a 'million to one' occurrence, it's still happening over 200 times a day. The sheer mass of activity on Twitter is its own system for amplification, and often not in positive directions. A study in 2016 (Hewitt et al., 2016) suggested that messages that could be legitimately defined as misogynistic came in at a rate of 6.6 abusive tweets per minute. The pattern of abuse too is not uniform, and overwhelmingly directed at particular target demographics. With the mass of aggression and unpleasantness online, expecting it to be a fertile ground for reasonable exploration of a nuanced situation is optimistic at best.

As a personal anecdote, one of the authors of this book was once branded as 'ableist and misogynist' by a Facebook user for giving prominence to issues of colour blindness on a website devoted to the accessibility of tabletop games. Social media removes the friction that exists in real-world interactions and permits both a kind of 'drive by' system of aggression and removes almost all of the social context that would moderate face-to-face discussions. Such a comment in the real world would be possible to discuss and dissect and would likely be delivered in a way less likely to instantly create friction. We are all more circumspect in our opinions when there are real-world social consequences that are linked to their expression. However, the nature of Facebook permits for such perceived 'truth bombs' to be dropped into a public space, detonated, and then for the source of the antagonism to disappear behind a shield of blocking and muting that prevents follow-up and meaningful discussion.

Many platforms permit for injudicious comments to be deleted or edited. That doesn't allow for commentators to undo the damage they may have done to themselves by interjecting an opinion into a fraught discussion space. Tweets and status updates may be screenshotted, and online services exist that archive tweets as they are made so that the fact of their removal or editing can be given undue prominence. Sometimes the act of recanting, or clarifying, will do little more than add fuel to a burning fire.

Individuals should enter into the arena of social media carefully because it is unforgiving of mistakes and missteps.

We can see this when Professor Blackbriar, frustrated by what he deems to be media complicity in a stitch-up, lets a momentary lapse in judgement potentially steer the course of the rest of his life. Presumably drunk, given the content of several of his blethers, he lashes out at Dunglen senior management accusing them of complicity and pressure with regard to publication of dubious data. The party previously seen as 'wronged' was likewise implicated in a conspiracy to mislead shareholders – ScotOil was alleged to have had a more active role than we have previously seen. It was claimed the Senate and Court of the university were directly involved in everything that was going on, and even his lawyers get something of a broadside when he acts against their explicit advice to 'keep quiet'. Blackbriar clearly believes that he is being slandered and that social media is the weapon he can use to clear his good name. In the cold light of day, six hours after the last tweet on the topic, Blackbriar claims that he has been hacked and his account is locked.

We have no way of knowing the truth of this, but it has to be said it rings hollow – and not just because it's so common a defence against social media misconduct that it has become cliched. Disconnecting from the discourse may perhaps be the only way Blackbriar can regain control of the situation – through actively sacrificing his own freedom of speech because of the dangers it carries with it. Certainly within the UK, the fact that a comment is made on social media does not inoculate its maker from legal sanction. UK blogger Jack Monroe won a case against Mail Online columnist Katie Hopkins over defamatory tweets. The judge found that there has been 'serious harm' to Monroe's reputation due to Hopkins' comments. American actor James Woods pursued but then dropped a legal action against a deceased Twitter user who had accused him of cocaine use. Sally Bercow arrived at an out-of-court settlement in a defamation case raised by Lord McAlpine. The judge ruled that her tweets were 'seriously defamatory' and legally indefensible.

The conversational nature of the medium, and the heavy role that interpretation plays in meaning (McCulloch, 2020), is not in itself a legal defence when circumstances go beyond that which can be settled with words alone. This is still largely uncharted territory as far as the courts are concerned, even given the way in which the incidents above have tested the legislative waters. These early indications suggest that there is a weight of consequence that must be taken into account when potentially libellous statements are made in a social media forum. We can see some evidence of this when actor JJ Welles was in 2023 forced to publicly apologise to JK Rowling over a tweet in which he labelled her a nazi.

With that in mind, it is perhaps not surprising that a measure of cold reality seeps in when someone has made a mess of a situation via social media interaction. Occasionally this can genuinely be attributed to hacking, but often the perceived anonymity of hackers serves merely as a convenient shield behind which wrong-doers can hide.

After his account posted a photo showing 'a real lamping' of 100 foxes, Vinnie Jones claimed 'this is a hack ive never seen this pic in my life and did NOT tweet it is a hack! !!'. Nadia Gustavo was a South African Zumba trainer whose account posted 'Friday 20, 2016, three black kaffas were finally found guilty of armed robbery in Queenstown regional court…'. On losing her job, she said, 'I have just lost my job at Virgin Active for something I did not do'. Amanda Bynes claimed she was hacked in a 2013 Twitter spat with rapper Kid Cudi. Anthony Weiner in 2011 initially blamed a mistakenly posted photograph of his penis on hackers, although he later recanted and apologised. C-SPAN host Steve Scully, who was lined up to moderate a 2020 presidential debate, made a public tweet that called into question his objectivity, and rather than accept responsibility he claimed that it was the result of hackers – something he had also done in 2013.

Sometimes it's not hackers who are claimed to be at work, but unnamed and unidentified assistants or staffers. Ted Cruz 'liked' a pornographic video on Twitter and claimed that it was the result of a staffing issue. This kind of response sometimes comes not even as a result of aggression, misconduct, embarrassment, or unpleasantness online. Alicia Keyes claimed she had been hacked after an innocuous tweet was posted from her iPhone – not a problem except she had just been made Creative Director for Blackberry. Sometimes the issue is merely that you've inadvertently outed yourself as being a supporter of an inconvenient brand.

We cannot be sure of the truth of such claims without a full investigation being conducted and the results being made available. It is though simultaneously plausible and implausible. Supposition and gut feeling must substitute for genuine insight into the real state of affairs. That is, of course, absent sufficient attention directed to the claim by parties able to uncover the truth and make it known. One might look to the 2019 'Wagatha Christie' incident when an argument between the partners of two footballers (Wives and Girlfriends, or WAGs, in the parlance of the

UK tabloids) blew up in the public discourse as a result of the investigative techniques of its wronged party.

In the case of the Scandal in Academia, not only are we unaware of the truth, we also know there is a motivated hacker working in the background to implicate major figures in the scenario. Claims of being hacked are often difficult to believe because of the convenience of their timing – they act as a kind of real-life deus ex machina that leaves people unconvinced. It is perhaps telling that immediately in the aftermath of the Blether comments our anonymous hacker revealed an email that substantiated many of the claims the professor's account had been making. Similarly, an anonymous insider at the company has given reasons to suspect there may be an element of truth to the accusations made against ScotOil itself. However, other emails leaked by Nemesis serve to act against the revelations made – contradicting the Professor's perhaps self-serving interpretation of events by releasing emails that put his claims under the light of scrutiny.

If he had been hacked by Nemesis, it seems unlikely the tweets would have run counter to the hacker's agenda. If he had been hacked by another anonymous hacker, how would they have known enough about the situation to make meaningful allegations regarding privileged and private information? On the other hand, perhaps there is an element of disinformation here – it doesn't do a hacker much good in this situation to make it obvious they're behind a faked Blether conversation, certainly not if their agenda requires the implication of a third party.

THE POST-TRUTH ERA

It's perhaps telling too that in the correspondence between Blackbriar and Sharon McAlpine we see the emergence of an argument that would later find prominence within the Trump candidacy for president. Blackbriar claims in an email, 'The criteria make no sense without access to the full data sets and would actually be misleading'. In 2016, Eric Trump in explanation of why his father would not release his tax returns, said, 'You would have a bunch of people who know nothing about taxes trying to look through and trying to come up with assumptions on something they know nothing about. It would be foolish to do'.

Whether this is a fair reason or not is up to each of us to decide as individuals. However, there is genuine danger that comes from non-experts reading too much into specialist data. Much of the climate change discussion, as we have outlined earlier, is very technical and couched in sophisticated internal jargon and convention. The sheer complexity of data can be a compelling reason as to why 'experts' should be ignored if one is pushing an advocacy agenda. Evidence denial (Lewandowsky et al., 2013) is a major reason behind the growth of pseudoscientific views in modern discourse. Having access to the data permits those that know enough to misinterpret it (knowingly or otherwise) to conveniently 'contextualise' it for a non-specialist audience. Most of us must rely on the analysis of experts in complex fields. We rely on doctors to interpret medical research. We rely on lawyers to interpret case law. We rely on teachers to interpret pedagogic theory. Even experts in one field cannot be relied upon to understand the context of another – those that work in health research cannot extrapolate the conventions that are germane to pedagogic research, and vice versa. It goes even further – we wouldn't expect an expert on human-computer interaction (HCI) to have much to contribute in terms of insight regarding the philosophy of dependence injection within software development… even if both topics fall underneath the general rubric of 'computing'.

We live in an age of decreasing trust of expertise and with this comes a growth in a mindset that believes facts are subject to interpretation. It is sometimes said we now live in a post-truth era. Claims such as these are part of the problem, simultaneously attempting to elevate one viewpoint as unquestionable truth and the other as 'post-truth' wish-fulfilment. When applied to raw, unquestioned data, this can be a fair comment – one is one, and that's not up for debate. When this kind of absolutism applied to complex, multivariate scenarios, this is something more likely to alienate people than convince them.

As scientific advocacy becomes increasingly politicised, we must accept the possibility that there are multiple interpretations for data and that an expert presents us with the one they find most credible. Ideally this is the one most compatible with the evidence, but it would be foolish to believe that sometimes it's the interpretation that is most compatible with an expert's politics or moral framework. It is sentiments such as these, and the occasional obvious example of experts or pundits pushing a biased agenda, that underlie claims such as that of Michael Gove when he said in 2016, during the national debate over Brexit, that 'Britain has had enough of experts'. This is a dangerous position for society to be in – when those most able to offer a clear and realistic lens onto data are held in the least esteem by those responsible for acting upon their advice.

That perhaps represents one of the most dangerous outcomes of the publish-or-perish mindset – that we're all so busy engineering the circumstances of publication that we don't have time to worry about the trust we may be eroding as a result.

THE DUNGLEN STITCHUP

What we can also see at play here is how additional pressure is being brought to bear upon the university through social media. Blackbriar's Blether outburst is only the start here – as a hashtag of the nature of something like DunglenStitchup starts to gain traction, it begins to generate a whole ecosystem of supporting 'content'. There are the initial messages, then the replies, then the counter-replies, then the quotes, then the Blether equivalent of 'tweetstorms'. Then there will be the memes, and the bots, and the doxing and everything else. The appetite of social media for scandal is limitless and Blackbriar has just rung the dinner bell and yelled 'bon Appetit'.

This represents an important challenge for everyone involved – whether Blackbriar was hacked or not, the fact is that this has progressed from being a contained incident of local importance to a hot topic online. It'll likely be picked up by Reddit, news aggregators, and cross-pollinate into other forums. The attention given to an incident like this might be short-term and burn out quickly. If it has enough juice in it, it might become a global preoccupation for weeks until everyone moves on to the next thing. They don't call it **trend**ing for no reason. There's a lot of fuel here though – it involves powerful white men being taken down; it involves the sacrifice of students, one a woman and another a disabled person; it has environmental aspects; it has implications of conspiracy. It adds evidence to the narrative of why you can't trust experts. There's a lot people can dig into. It's only to be expected that YewChoob will soon be full of video essays from people who previously had never heard of Dunglen and yet are now seemingly experts on how the university works. A thick miasma of information, misinformation, and disinformation is bound to emerge. Hot takes. Unsourced gossip. Welcome to the modern media circus, and Blackbriar just made himself chief clown. Those working at the university now need to be thinking about how to manage the public perception of the scandal, which is unfortunate because at this point it's unlikely the beast can even be steered, much less tamed.

The Dunglen Chronicle, on the other hand, is undoubtedly absolutely thrilled. A local newspaper suddenly coming to global prominence is going to make a lot of people very happy. Those with a financial stake, sure – those clicks will turn into ad revenue, and the online discourse will probably sell a few more papers as well as drive spend on the site. Those like Jack McKracken who will find themselves with a global audience eager for more revelations. This can be a career defining moment for a journalist – when obscurity finally gives way to name recognition. We're going to see – assuming again that this story has some legs beneath it – growing factionalism. There will be those that believe the stitch-up narrative. Others that are just delighted to see someone in authority being 'held accountable' for his actions. There will be self-appointed social defenders and armchair prosecutors of everyone who is mentioned. The internet, as a collective entity, will turn its archivists to digging up more details and uncovering ever more invasive information. Did Blackbriar ever have a relationship with a student? Who is McAlpine's husband and why is she still with him if she's so obviously unhappy?

Just exactly how blind is James Duncan, and to what extent can be speak for the blind community?

These questions do not belong within the remit of this scandal, but that's not going to stop people. Consider Charles Ramsey, who in 2013 assisted in the rescue of three kidnapped women. He became a local celebrity as a result of his actions. The attention led to public scrutiny of his life. This revealed that he had been arrested several times in the past for domestic abuse. Marcus Hutchins stopped the Wannacry virus in its tracks during 2017. He did so under a pseudonym. Public speculation was so intense that British tabloids doxed him and revealed his name along with unrelated biographical details to capitalise on the money that comes with clicks on their website. In 2022, anti-Covid vaccination activists drove Lisa-Maria Kellermayr to suicide after a prolonged online harassment campaign. Her initial crime was to send a tweet criticising protesters. A brief candle of internet celebrity – especially if it comes with the frisson of hidden truth or controversy – can result in someone being publicly destroyed for unrelated activities or from being exposed to the heat associated with the simple flame of attention.

Aside from pointing out a few times that all the reporting on the scandal comes from a single newspaper, we've left Jack McKracken largely uncritiqued. Now perhaps we should be paying more attention to the way in which he has been framing the discussion because he's just been given a powerful incentive to keep eyeballs on his articles. We might assume journalistic ethics ensures that he isn't saying anything untrue, but there's a difference between the truth and the **whole truth**.

We should be critical here – is he reporting all sides of the story equally? Is he quoting people out of context? Is he choosing the topic to cover based on what contributed to the public right to know, or that which is most likely to sell papers? To be fair, it's possible for him to be serving his career while still being largely in alignment with a broader ethical code… but the fact is that this is the only lens we have on what has happened, and we don't know what impurities that lens may possess. Jack McKracken, previously the custodian of the story, is now going to find himself just one voice amongst many – perhaps the one with the greatest authority by virtue of having made the largest contribution to the story being uncovered. Like everyone else involved though, he better hope his history withstands scrutiny.

More than this, the social media response itself is going to become a story – it will generate other stories in other outlets purely because of the attention they gain. It's not that the story has become more newsworthy in and of itself, but rather the media is complicit in feeding the fires of social media outcry. What we'll see coming from this is a range of 'meta-commentaries'. We'll see articles that capture what people are saying online. We'll see academics writing papers about it. We'll see television news try to turn flowing screens of text into video content by layering tools like sentiment analysis onto the discussion. We've just seen that in our newest article.

That in turn is only going to skew perspectives, since the specific flavour of sentiment analysis used will dramatically change how comments are coded (Shen et al., 2018). Use one sentiment analysis algorithm and perhaps 60% of messages are supportive of Blackbriar. Another may report only 30%. Results can vary by language, depth of subtext, specific cultural jargon, and other complex factors. The choice of algorithm used in such circumstances is rarely transparently presented to a viewer or a reader. Despite this opaqueness, these commentary pieces will shape the public response even as the public response shapes these commentary pieces. Proverbs 27:17 says, 'As iron sharpens iron, so a person sharpens another'. That's what's going on here. As Lenin once said, 'everything is connected to everything else' and nowhere is that more true in how people engage in discourse online.

The problem with the kind of analysis that uses algorithms to produce an illusion of objectivity is the same issue we discussed in the previous chapter with regard to metrics. Analysis can never reflect the true depth of nuance to be found in any issue while still counting as analysis. By its nature, it must attempt to identify themes and meaning so as to convert information into knowledge. In the process, important things might go missing. Scrutiny may be applied unevenly. The role of analysis in the construction of a narrative from complex events is sometimes to cut away at those elements seen as extraneous. Those elements may turn out to have been important all along.

A quote often mis-attributed to Mark Twain is 'If you don't read a newspaper, you're uninformed. If you do read a newspaper, you're misinformed'. Newspapers encode the biases of their readers, and as such each different agenda results in a different flavour of misinformation. The more we see external media shaping the discussion, the more noise we're going to see interfering with the signal. There is no opinion so evidence-free, so bizarre, and so extreme that you can't get about 5% of the internet to agree with it. There will be dark corners of Reddit that assign all the blame to Sharon McAlpine. There will be social justice vigilantes that just want to see Blackbriar burn because he happens to be a middle-aged white man. There will be well-meaning but ultimately patronising commentators who will assume James Duncan is guiltless because it is ableist to believe someone with a disability can also bear culpability. The further you dig into the social media response, the wilder and more alarming the commentary will be. The more alarming the commentary is, the more people will read the articles written about it.

The University of Dunglen wanted to escape this issue with its reputation mostly intact. Those affected likely wanted to resolve the issue privately so that it didn't destroy their public reputation. One suspects nobody – except perhaps Karan Chandra, who has taken centre stage in a lot of the rhetoric – is happy to find out they're now the main attraction in an online carnival.

REFERENCES

Hansen, L. K., Arvidsson, A., Nielsen, F. A., Colleoni, E., & Etter, M. (2011). Good friends, bad news-affect and virality in Twitter. In *Future information technology* (pp. 34–43). Springer.

Heron, M. J., Belford, P., & Goker, A. (2014). Sexism in the circuitry: Female participation in male-dominated popular computer culture. *ACM SIGCAS Computers and Society, 44*(4), 18–29.

Hewitt, S., Tiropanis, T., & Bokhove, C. (2016). The problem of identifying misogynist language on twitter (and other online social spaces). In *Proceedings of the 8th ACM Conference on Web Science (WebSci '16)* (pp. 333–335). Association for Computing Machinery, New York, NY. https://doi.org/10.1145/2908131.2908183

Lee, E.-J. (2007). Deindividuation effects on group polarization in computer-mediated communication: The role of group identification, public-self-awareness, and perceived argument quality. *Journal of Communication, 57*(2), 385–403.

Lewandowsky, S., Oberauer, K., & Gignac, G. E. (2013). NASA faked the moon landing—Therefore, (climate) science is a hoax: An anatomy of the motivated rejection of science. *Psychological Science, 24*(5), 622–633.

McCulloch, G. (2020). *Because internet: Understanding the new rules of language.* Penguin.

Ronson, J. (2015). How one stupid tweet blew up Justine Sacco's life. *New York Times, 12.*

Shen, J. H., Fratamico, L., Rahwan, I., & Rush, A. M. (2018). Darling or babygirl? Investigating stylistic bias in sentiment analysis. In *Proceedings of the Workshop on Fairness, Accountability, and Transparency (FAT/ML)*, Stockholm, Sweden.

Williams, K. R., & Guerra, N. G. (2007). Prevalence and predictors of internet bullying. *Journal of Adolescent Health, 41*(6), S14–S21.

CHAPTER 11

Culture of Fear and Nepotism at University

NEWSPAPER ARTICLE

An exclusive report by Jack McKracken for the Dunglen Chronicle

Students today claimed that the University of Dunglen campus has been enveloped in a growing 'culture of fear' as the university senior management team attempt to handle the crisis at the centre of the institution.

'People are afraid to say anything', says one student speaking on condition of anonymity. 'We all know that this is a big deal and that our own degrees are in danger of being worthless by the time we graduate. I'm seeing so much on social media, and it's a frenzy. There are all kinds of conspiracy theories floating around too but it's like Soviet Russia – it's all whispered in dark corridors and everyone scatters when a stranger walks by'.

A second anonymous student said, 'I look around and all I see are worried faces. I didn't realise how important this was until I saw how other people were responding. But now I understand – if the University closes, or loses its way down the league tables by the time I graduate, nobody will remember where it started off. I get the reputation the university has when I look for a job, and that has a big impact on me'.

'Last night I was looking up some stuff in one of the computer labs', a third student tells us. 'It was about the vice-chancellor and his past work in the oil industry. A guy wearing a university lanyard came in, saw what I was doing, and insisted that I close down all my browser

DOI: 10.1201/9781003426172-11

windows. And I did, because I don't think it's a good time to get on the wrong side of anyone in the staff at the moment'.

'It's really ruined the entire semester', said a fourth student. 'University is a lot of hard work, but it's also been a lot of fun up until now. We went out drinking after exams, had parties at weekends, it was all great. Especially after we all got locked up in dorms during the pandemic, we really need to enjoy the people around us. Now though everyone is afraid to do anything to stand out from anyone else. There are people with cameras on campus every day – stopping people at random to ask them questions or to scream in their face. You know that an hour later it'll all be edited and put on someone's socials. You might find yourself overnight the target of trolls and cyberbullies just for giving an opinion or reacting poorly under stress. Nobody wants to make a joke, tell a story, or call attention to themselves in case we end up as part of tomorrow's news'.

The punitive atmosphere that has taken hold of the University will not have been helped by the most recent revelations of the hacker known as Nemesis. Secret emails sent between senior members of the university management and Professor Blackbriar show some of the private comments made about individuals applying for the PhD studentships with the professor. The first of these relates to the applicants directly:

To: John.Blackbriar@dunglen.ac.uk
From: Ian.McManus@dunglen.ac.uk
Subject: PhD Applicants

Hi John

Sorry you didn't have time to really dig into the applicants, but I think the selection committee you set up has done a good job.

They're all very capable, as we might expect given the project they'll be working on. We had twenty-eight applicants for the two positions. That is, in terms of the ones that made it through the initial AI sift. Two-hundred and thirty-six candidates before that. Healthy numbers, but that's not a surprise. Who wouldn't want to work with the great John Blackbriar?

We've whittled down the short-list to four. We've tentatively set the date for interview to the 14th of September – are you free then? We'll do the normal thing – tour of the campus, presentation, then interview. Here's a run-down of my initial thoughts on the ones we've selected.

Sharon McAlpine is the only female applicant –

just as qualified as any of them, but I'd say we should put a pin in her if she does well at interview. The gender balance in the computing department is pretty bad, and she'd do wonders in redressing that. I'm not saying 'Definitely her', just you know – if we need a tie breaker. She's also very pretty – she'd certainly brighten up that drab office you keep your students in!

She does have a new kid though, and I know how much you value the availability of your students.

Dirk Tumblewood is the VC's nephew. An arrogant little shit, but he's got credentials coming out of his arse. Double firsts from Oxford, two years of interning at BP, and a list of contacts that even you'd be jealous of. I will warn you though that he's been picked up a few times for using chatbots to write parts of his master's assignments, but he's been behaving well over the past year.It's not on his official transcript, nobody wants to piss off the VC with an inquiry. You know how it is.

James Duncan is almost entirely blind, although still a dab hand at coding. If you want to hire him, he'll need some special arrangements for his dog and for suitable accessibility equipment to be installed on his system. There might also be an issue with accessing some of our internal programs – some of them are a bit creaky and I don't know if they work with screen-readers or whatever. That's all logistics stuff, and I think it might make him a fair bit more expensive and troublesome to hire than your average research student.

His references all say that he's shit-hot at the 'computing stuff' though which is just what you said you need. I'm not saying not to hire him, of course, just be aware of the budget considerations. Finally, Stan Templemore – seems competent, ticks all the right boxes, but it's hard to get excited about him. My bet is that he'd meet all your requirements but you'd be hard pressed to pick him out of a lineup of one after three years. He's really well liked though – spends a lot of his time supporting his colleagues rather than building his own career. I'm sure you'd knock that out of him though – you always have before.

My gut feeling, pre-interview, is that McAlpine and Tumblewood would be the most promising candidates. You don't want to get on the wrong side of the principal, I'm betting!

Talk to you later, Ian.

The university was quick to respond to the suggestion of nepotism on the part of the principal, saying, 'This is a joke clearly made in context between individuals who know each other well. There was absolutely no pressure applied by the principal in the selection of candidates for the position, as Professor Sir David Tumblewood well understands the importance of avoiding any conflict of interest even in staffing PhD studentships. He wasn't part of the selection committee and played no part in the interviews'.

The University's account of this may have been borne out by the second email released today, which is dated some time after Sharon McAlpine and James Duncan were offered the positions:

To: Ian.McManus@dunglen.ac.uk
From: David.Tumblewood@dunglen.ac.uk
Subject: Dirk

I'm pretty disappointed that John passed on Dirk's application. I've had his mother on the phone to me all day. Not surprised he went for the lady – he's really got an eye for the talent, as it were.Does he have a backup plan for if she gets pregnant again? We really can't afford the delay of maternity leave.

Very surprised about the blind kid – I guess it does our equality profile good though. Did John motivate why that was his pick?

Did he give a reason why he rejected Dirk?I know he's a pain in the arse but he's a good lad really. It's pretty rare that fully funded PhDs come up in this area, and he had his heart set on it. Oh well. I'm sure John knows what he's doing. Maybe we'll get lucky and one (or both) of the ones he's picked will refuse the post. We might find room for Dirk then.

See you tonight at the reception, David.

Both Sharon McAlpine and James Duncan have responded with anger to the leaked emails. Mrs. McAlpine issued the following statement via her lawyer, Karan Chandra:

'I am saddened and upset, but not surprised, to see the leaked emails today. My qualifications as an outstanding candidate for a postgraduate research position were overlooked in favour of patronising comments about my "brightening up the place" and meeting a gender equality benchmark. It is precisely this kind of casual misogyny that creates an environment where women don't want to participate. The best solution to fixing the issue of a lack of women in technology is for opinions like this to go the way of the dinosaurs that spout them'.

Mr James Duncan's statement takes a similar tone:

> I'm not just an entry on a diversity spreadsheet, and it is unconscionable that the University felt that the cost of making basic provisions for the blind should count as a reason for not hiring me. I am grateful that Professor Blackbriar, for all his faults, picked me for my capability and ignored my disability.

Karan Chandra issued his own statement in a press release, saying:

> It is clear from the vice-chancellor's hopeful tone in his correspondence that he was looking for a reason to kick both Mrs. McAlpine and Mr. Duncan off of their doctoral programs so that he could clear the way for his nephew. Suddenly the reason why my clients were so quickly and abruptly thrown to the wolves becomes clear.

'None of this makes me feel any more comfortable about the university', says our first anonymous student. 'Are they talking about the rest of us like that when they send emails around? What are they saying about me when I ask for help? Do we all really need to worry when the principal is on the warpath in case we get on his wrong side? It just adds a pile of stress to an already stressful situation. I can't believe I'm paying tuition fees to study in what seems like a stalag. It makes me yearn for the pandemic at least then I could study from the comfort of my bedroom and I didn't have to sink into the sadness of the physical campus. They don't let us do remote learning any more though'.

Professor Blackbriar last night was unavailable for comment.

FEAR AND NEPOTISM AT THE UNIVERSITY OF DUNGLEN

We should begin this examination of the issues in the Scandal in Academia by addressing the atmosphere at the university, since the social context in which everything is occurring is something that is powerfully influential on the way everyone is behaving. There is much in the way of implication in the statements made by the anonymous students that introduce the article. Particularly stressed are fears:

1. Fear their degrees will be devalued
2. Fear to talk freely due to the suspicion journalists walk amongst them

3. Fear that a secret gossip circle exists in professional emails
4. Fear that free inquiry is being suppressed

These are not unfounded. The value of a degree is adjusted by the perception of worth that others ascribe to it (Donald et al., 2019; Miller et al., 2015) particularly when it comes to job-seeking and career-building. A degree from a high-ranking university is often a prerequisite to be taken seriously as a candidate for some jobs (Grugulis & Stoyanova, 2012). The disproportionate hold that elite universities and private schools have over some professions (Davies, 2018; Friedman et al., 2019; Wai & Perina, 2018) is a powerful indicator of the social cachet they confer. In more mundane circumstances, we see pressures on students during evaluation exercises such as the National Student Survey to suppress negativity to ensure that the university doesn't drop down league tables, something we have addressed in previous chapters.

A degree is a mark of social proofing and if that proof is seen to have come from a tainted source it will not have the same value as it would otherwise. Keeping quiet is a self-preservation tactic in the short term even if it carries with it considerable long-term disadvantages for individuals and the institution itself (Fanghanel & John, 2015). Given the way our scandal is unfolding, it is reasonable that students would not want to contribute to a situation where every future job interview contains the question 'You went to Dunglen – I heard there was a big problem there during your studies, what can you tell me about that?'. Given the circumstances and the amount of scrutiny the university is under, it is unlikely that there are any formal or official pressures on students to avoid being candid. Self-interest is enough to encourage discretion.

Given the suspected presence of influencers and journalists at the university, it is likely that a degree of ingroup versus outgroup thinking has begun to take hold (Abrams et al., 1990). Policies regarding involvement with the press are inconsistent between institutions and even where they are well formed and unambiguous there is often a question mark on how rigidly they can apply to students as opposed to staff. For most members of an academic community, this is a small consideration – few become prominent enough to attract the attention of journalists. More commonly, universities and other institutions make use of social media monitoring to track individual commentary, and there are often policies that cover the acceptable use of social media (Pasquini & Evangelopoulos, 2017; Pomerantz et al., 2015). While institutions lack the ability to censor or control all social media use, some exert specific prohibitions on students when they talk about the university. Increasingly, surveillance is a cost of receiving an education (Beetham et al., 2022). Nottingham university, at the time of writing, has the following provision in their 'Social Media Policy for Students'[1]:

> This policy applies to all students and to any personal communications within a social media platform which directly or indirectly reference the University.

And:

> The University acknowledges that students may use social media in their private lives and for personal communications. Personal communications are those made on, or from, a private social media account, such as a personal page on Facebook or a personal blog. In all cases where a private social media account is used which identifies the University, it must be made clear that the account holder is a student at the University to avoid the impression that views expressed on or through that social media account are made on behalf of the University

In all cases though, the ability for institutions to effectively police any specific group is limited to the visibility enabled by individual social media platforms. There is evidence to suggest though that these policies are not truly about ensuring the best experience for students and tend to emphasise the rights of an institution to control its own reputation online (McNeill, 2012). Social media is increasingly seen as a branch of marketing for a university (Constantinides & Zinck Stagno, 2011; Luo et al., 2013; Rutter et al., 2016) and its staff (Donelan, 2016) and thus reputation protection and brand management is often the primary driver for these policies. Indeed, when social media activity is deleterious to the brand, more established principles such as freedom of academic expression will inevitably suffer. As Ringmar (2007) says, 'The limits to freedom of speech are set by the market'. Derek Morrison, of the UK's Higher Education Authority (HEA), quoted in McNeill (2012) says bluntly:

> The simple rule for everyone should be don't affect the share price, no matter what technology you are using.

This is relevant here because social media is a primary tool by which journalists can get access to critical information when investigating an institution of any kind (Abdenour, 2017; Gearing, 2014). The often porous nature of the staff/student interface in a university makes this a particularly effective technique. This is an environment full of academics with expertise in non-technical disciplines but limited awareness or experience with online disclosure (Masur, 2019), especially in the face of the 'privacy dark patterns' employed to keep data flowing (Bösch et al., 2016). This combines with a general apathy about disclosure online amongst younger users (Hargittai & Marwick, 2016). These and other factors mean that journalists can often get access to greater insights than they might otherwise in a locked down information ecosystem. It's difficult for a journalist to just walk onto a Google campus, for example.

However, the older tools of investigative journalism are just as effective when applied to a university as they are elsewhere – asking questions, making phone calls, and simply hanging around listening to what others say. Permitted presence on campus is technically regulated by an ID card of some kind, but enforcement is often lax except under special circumstances. In most of the institutions where the authors have worked, staff cards have generally remained in wallets and the number of times this has resulted in a challenge during the course of regular duties is in single digits over two decades. On one memorable occasion, one of the authors of this book – dressed in jeans and a t-shirt – needed access to a locked room full of expensive specialist equipment. It was an unfamiliar building in an unfamiliar part of the campus, but simply going up to the front desk and asking for the key was enough to get access. Act like you belong, and people often won't stop to ask if you do. Such is the core of social engineering.

Security practices may vary from institution to institution, and even from year to year. Post-pandemic, the university where the authors currently work has been described, by Michael, as 'like working in a CIA blacksite'. The number of card-swipes required to get into the building, into an office, and then into a classroom is genuinely bizarre. Much of this is a pantomime of security – access to the bridge between buildings for example requires a card swipe, but the two different buildings can be accessed without a card by simply by going outside and entering via the main door.

Sometimes external events influence the theatre of security – when a member of the royal family was visiting a different institution, staff were denied access to the building and surrounding complex if they could not produce appropriate evidence. They were queried even if they did as to their purpose – at a building where they were employed. This inconsistency of enforcement in itself is a tool that journalists can leverage should they be so inclined. Sometimes they don't even need to leverage anything – all of this is leaving aside the occasional incidents where leaks are used as a tool in winning internal struggles within a university[2]. The press is a proxy battleground where arguments of all kinds can be fought and occasionally won.

In times of scandal, such as we see here, organisations will often take the opportunity to 'remind' people of convenient policies that were never effectively communicated in the first place. Temporary prohibitions may apply, and special instruction may be given. For example, during a particularly problematic incident involving the alleged misconduct of a vice chancellor[3], staff at the affected university were given explicit instructions about how to identify journalists and how to refer them to the university's PR department in line with a policy that had hitherto been under-emphasised. Policies for social media for staff are inevitably stricter than those for students, but the same general principles apply.

However, the simple presence of people with cameras on campus may be enough by itself to create a chilling atmosphere for those that are simply looking to go about their studies. Professional journalists can be expected to conform – at least in principle – to certain norms of conduct. Even with that, breaches of privacy by journalists are often shocking in their brazenness. We've already briefly mentioned Marcus Huchins, the cybersecurity specialist who stopped the Wannacry virus by identifying its weak spot and creating what was effectively a kill switch (Zimba & Mulenga, 2018). This he did under the pseudonym of MalwareTech. Speaking to the Guardian, he indicated that he wished to remain anonymous[4]:

> It just doesn't make sense to give out my personal information, obviously we're working against bad guys and they're not going to be happy about this.

His right to privacy was not respected. Within days, The Sun, The Daily Telegraph, The Daily Mail, and the Mirror had worked to find out his name. They published photographs, camped out on his lawn, and hunted down colleagues for insights into the reclusive young hero. In other incidents, journalists associated with News International have 'hacked' into voice-mails to get access to information for stories. Whether this is acceptable under any circumstance is a matter for debate – and in many cases, litigation. However, these activities were not confined to celebrities or figures associated with the public's ill-defined 'right to know'. Investigations carried out as a result of police inquiries revealed that journalists had been using these techniques to access the voicemails of people associated with the murdered schoolgirl Milly Dowler. Also of the missing Madeleine McCann, and the family of the Sohan Children. There is also evidence of the same being applied to relatives of deceased British Soldiers as well as victims of terrorist incidents including 9/11 and the London bombings. Members of Parliament (including Tony Blair, David Blunkett, Boris Johnson, and Gordon Brown) and their relatives (Cherie Blair, David Mills, and Gaynor Regan) have been targeted. Members of the legal profession were considered fair game (Kirsty Brimelow, David Cook, Lord Imbert, and Michael Mansfield) as were witnesses in legal proceedings (Anne Colvin and unnamed witnesses to the murder of Jill Dando). Christopher Jeffries, wrongly accused of the murder of Joanna Yates, had his messages accessed while he was being defamed by the newspapers doing the defaming. The list of victims of this invasive technique is… extensive[5].

If this is what professional journalists will do in pursuit of a story, imagine what restrictions a budding influencer may place upon themselves in the war for attention on the Internet. The average age of an employed journalist is around 40 years old, with the associated expectation of life experience and maturity… despite the evidence outlined above. The average age of a 'successful' influencer is 27 years old. The average age of a budding influencer of the kind most desperate to stand out from the crowd is much lower, with some estimates suggesting a median age of around 19 years. What expectations should we have on a teenager arriving on site in the middle of a complex situation? Especially when their entire motivation is to build a public profile for themselves that will allow them to rise above competition? What price will someone be willing to pay for virality?

Well, we have some benchmarks. In December of 2017, successful influencer Logal Paul entered the Aokigahara Forest, famed for being a home to ghosts and the site of many suicides since the 1960s. There he filmed, and then published, a video in which he recorded a dead body and his reaction to the same. He was 23 years old at the time. Nicole Arbour in 2015 posted a video widely considered to be 'fat shaming', which went viral to great controversy. She later revealed the whole thing – the controversy and her later climbdown – was a marketing scheme worth tens of thousands of dollars. PewDiePie, at one point the largest star on YouTube, released a video in which he hired people on the crowd-labour site Fiverr to display a message saying 'DEATH TO ALL JEWS'. The list of incidents like this is vast and growing – success through engineering virality at the cost of authenticity is a common technique for those hoping to turn attention into a career. If you need more evidence, we might point to ImJayStation who faked his girlfriend's death to get more subscribers – blatantly and intentionally using his conspicuous artificial grief to garner a new audience on the grounds 'it's what she would have wanted'. YouTuber ReSet received a 15-month custodial sentence for giving a homeless man an Oreo cookie filled with toothpaste and recording the result. Sam Pepper faked a kidnapping. DaddyOfFive and MommyOfFive lost custody of their children after abusing them online in exchange for YouTube views. Honestly, the list goes on and on.

For every significant case of outraged notoriety, there are many more that simply went unnoticed because the uploader was not famous enough to garner real attention. The Internet is full of pranks being performed on innocent bystanders, or sexual harassment of women passed off as comedy. No one knows the formula for success as an influencer, but **recency** is undoubtedly a major factor. It's hard to become a star when you take your time to have a slow, measured, and thoughtful response to an incident that happened in the past. It's much easier to throw out 'spicy hot takes' and 'spill the tea' – two terms that refer, respectively, to speaking your mind without allowing opinions time to evolve, and to reveal gossip with the intent of titillating a receptive party.

It's obviously unfair to judge all journalists and all influencers by the actions of the worst of them, in the same way we can't judge all academics by the standards set by the frauds we have discussed in earlier chapters. There are people out there doing wonderful, uplifting work. There are video essays with all the density and robustness of the best academic papers. However, that's not the way you want to bet it in the early days of an emerging scandal. Influencers inevitably benefit from the parasocial relationship their viewers have with them – that is to say, a one-sided arrangement where one person looms large in the mind of the other, and the other has no idea the first exists. The personalities shown by influencers online are crafted through curation and editing in the same way that data sets are improved through pruning 'outliers'. Everyone is putting forward their best face, in the best way. Authenticity is difficult to judge at a distance. Every interaction with an unknown journalist or unknown influencer puts someone in the shoes of the imprisoned within the Prisoner's

Dilemma – if you both trust each other, you will both benefit. However, the temptation to screw over the other is hard to resist (Rapoport, 1989) because the rewards can be so much greater. Students may be wary of middle-aged journalists, identifying their 'Hello fellow kids' rhetoric as inherently inauthentic. It'll be harder to spot the nefarious influencers given how close they'll be in age group.

University staff may find themselves targeted specifically. They have more access to information on what's going on, and that information tends to be at a level of greater proximity to that of the decision makers in a university. University staff as a whole though tend to have more experience in dealing with difficult questions and allowing themselves time to consider before responding. The risks to them are also correspondingly greater – most will have contractual obligations to avoid inflicting reputational damage on their employers. As a result, caution is to be expected on their part. Even those that have never talked to a journalist can often rely on the reflexive skills that come with a career of giving lectures, discussing complex issues in tutorials, and publishing academic papers. Journalists and influencers want quick, easy soundbites they can edit into viewer-friendly packages. They tend not to be so keen on the rambling 'on the one hand this, on the other hand that' approach taken by your average scholar.

Nonetheless, one cannot discount the power of social pressure. Intention can be secondary to simply responding to a lifetime of conditioning. The shock of being ambushed and put on the spot can result in people – especially those without media training – disclosing damaging information without ever intending to (Keller, 2019). Even simply being subject to ambush interviewing can make people look guilty, and viewers will interpret this accordingly (O'Neill, 1983). The risks go even deeper than that – direct participation is not required to reveal compromising facts. Simply being overheard in an unguarded moment may result in damaging information suddenly entering the public discourse.

Staff and students at the university undoubtedly have reason to be concerned about the sudden influx of strangers wielding selfie-sticks and cameras. They're right to be concerned that people may be listening to what they say. Combined with the potential market damage to the 'share price' of their degree, a culture of paranoia is not a surprising outcome. This paranoia is fed too by the suggestion in the leaked emails from Nemesis that members of staff are gossiping about people in a way that doesn't exactly meet the general standards for 'professional discourse'. People being people, it's inevitable that this should be true.

On one level, it's good practice for staff to cross-talk amongst themselves about individual students – it permits the identification of problems (Graham et al., 2016), facilitates interventions at the earliest opportunities, and generally permits important information to permeate through a faculty. Knowing a student has had a bereavement, even if it is through gossip and back-channels, can be the thing that encourages a lecturer to be additionally and mindfully supportive during an interaction. That in turn can be exactly what a fragile student needs at the time. The authors can say that most of the 'gossip' we have seen as academics is positive or supportive – making sure student accomplishments are made known within a department or ensuring the circumstances are known for enhanced pastoral consideration.

It certainly can't be argued that all gossip is noble though. Its role as a facilitator of abuse and harassment is well studied (Branch et al., 2013). Academics are as predisposed to vent as anyone else. Frustrating interactions with students may result in fiery email chains or energetic discussion in the staff room. Occasionally these emails make their way into student hands, such as by a lecturer with a particular student on their mind emailing that student directly instead of the colleague that was intended. That may sound unlikely, but as the kids say these days 'We have seen the receipts'. The impact such accidental indiscretion has on the relationship economy upon which education is built is considerable.

Perhaps the most worrying of the concerns students have is that their ability to freely investigate issues is being suppressed by members of the university staff. Valid, and perhaps entirely innocent, investigation of staff members prominent in the university is alleged to have been shut down. The implication there is that academic freedom is being suppressed through monitoring and surveillance of student activities. This is likely something counter to the university's own policies regarding acceptable use of university resources as well as the university's legal requirement to support an environment in which academic freedom is protected. However, the latter of these is something that has always, in practice, been secondary to the reflexive instinct of a university to protect itself and its brand (Christensen et al., 2019; Giroux, 2006; Lieberwitz, 2002; McLelland, 2019; Wilson, 2015).

Citing the existence of a policy by itself is often enough to convince people they are in violation given how such policies are often baroquely worded, ambiguously applicable, and implicative of dubiously justifiable penalties and dire costs for violation. Social pressures though are powerful, and when instructions to refrain from an activity come from someone in a position of authority – even one as incidental as is bestowed by a lanyard – it can become an issue of over-application. Consider the obvious levels of stress and concern expressed throughout the university at the emerging scandal. To fully understand the atmosphere, we need to decide to what extent students feel empowered to defend the rights they have been given. Especially given those rights are enshrined in a matrix of legal protections, policy specifics, and informal conventions around the academy. Can we simply assume that students in general even know the broad outline of their academic rights?

Important though here in the case study is that these are anonymous comments and thus we have no way as readers to ascertain the validity of the source or the extent to which their comments represent outlier or mainstream perspectives within the institution. We don't even know if these comments were provided first hand, second hand, or via an observation of social media discussion. Much of the information we receive throughout our case study must be assessed in this light. We simply have no way of calibrating the extent to which these views represent a consensus of the student population. We can say though that there are believable real-world parallels for why students may be seeking shelter in anonymity at this time.

SYSTEMIC MONITORING OF STUDENTS AND STAFF

We should return to the issue of student monitoring because in some countries it is now a formal part of a university's legal functioning. In the UK, universities are legally designated as 'specified authorities' under the Counter-Terrorism and Security Act (CTSA) – officially known as PREVENT. They thus must, 'in the exercise of their functions, have due regard to the need to prevent terrorism'[6].

What this amounts to is that staff must be furnished with formal training designed to help them identify signs of 'radicalisation' on the campus. Staff have a duty to then report incidences that could be interpreted as leading students down a path that would endorse terrorism or other acts of violence. This is now a contractual requirement of every academic in the UK, with consequent impact on employment if these duties are not carried out. This has, understandably, created an energetic debate about the role universities have in policing acts of speech and the extent to which universities as an institution can ethically designate topics as being off limits for expression. C.f. (Bryan, 2017; Gilmore, 2017; Greer & Bell, 2018; Spiller et al., 2018).

That particular debate is unlikely to reach a conclusion any time soon, and the authors of this book will refrain from commenting one way or the other. Instead, we simply note that this legislation requires a form of systemic monitoring and reporting on certain student activities in a way that can certainly be seen as problematic. As Gearon and Parsons (2019) argue – there is a grey line where reasonable scrutiny becomes unreasonable surveillance. In particular, objections have been raised arguing:

1. The specific flavour of guidance provided on the topic is either implicitly Islamophobic or so general as to be useless.
2. Academic staff are not sufficiently trained in issues of counter-terrorism to be able to effectively and correctly identify inciting incidents.
3. Implicit biases derived from political sentiment and media coverage mean that PREVENT-based interventions are disproportionately based on profiling.

The consequences to an individual academic can be significant. The guidance at the University of Sussex outlines as follows[7]:

> Our policies place a considerable degree of responsibility on you to assess the risks of any event that you organise. In the overwhelming majority of cases, the risk will be negligible and so no further action will need to be taken. However, it is essential that you take this seriously and a failure to comply with your duties will be treated as such. If you aren't sure, you should consult with your Line Manager.

Note here the guideline that 'the risk will be negligible in the overwhelming majority of cases' combined with the note that there are disciplinary issues associated when it comes to interpreting often conflicting guidance. PREVENT training is sometimes notable for how unspecific it is. When given criteria by which staff might judge whether someone was likely to be demonstrating cause for concern, guidance included suggestions to look for:

1. A change in friend groups
2. Support for causes likely to bring around significant political changes
3. Changes in attitudes or viewpoints

One might note that these criteria apply to a vast majority of undergraduate students at one point or another. Universities were once lauded as one of the places where students do become radicalised to positive change – the genesis or popularisation of a lot of popular protest movements is historically within the university campus or through affiliated youth groups (Heineman, 1992; Milkman, 2017). If we are not educating students to change their viewpoints in the face of evidence, we are failing in our basic duty as educators. The broad and unspecific nature of this training just as easily covers being a radical proponent for free tuition as it does someone courting with the principles of domestic terrorism. It was explained later that this framing was to ensure that the guidance didn't racially profile or religiously discriminate. The cost of that vagueness is that every student is a potential radical and university academics are thus required to report on their behaviour where it is deemed to transgress some invisible and ill-defined threshold of extremity. Training experienced by the authors of this book was anecdotal and purposefully vague, which made it especially chilling in its implication. Students that feel overly monitored may be justified in their disquiet at this kind of legislation and how it has worked its way into the academy.

At the same time, academics do have a duty of care and were similar guidance to be issued on identifying issues of social isolation, risks of suicide, or harassment it would be unconvincing to argue that this presents a risk to freedom of expression. It is a difficult argument to make that universities alone should be exempt from counter-terrorism policy by virtue of their largely self-proclaimed status as havens of free inquiry. However, legitimate arguments exist about where the balance lies between coherently addressing issues of violent radicalisation and permitting the 'expression of ideas without risk of official interference or professional disadvantage'.

It would lack nuance for this book to argue that authority and constraints on freedom of expression come only from top to bottom. While PREVENT and related legislation does apply to staff and the monitoring of student opinions, there is a growing trend of 'no-platforming' taking hold in academic discourse. Staff inviting external speakers on a topic may have no PREVENT concerns. Student advocacy though has grown increasingly strident in recent years regarding the practice of denying controversial voices the right to speak. The idea was expressed by the National Union of Students (NUS) thusly:

> The policy prevents individuals or groups known to hold racist or fascist views from speaking at NUS events. It also ensures that

> NUS officers will not share a public platform with individuals or groups known to hold racist or fascist views

The six organisations specifically targeted by this policy, as of its last update, were:

> The six organisations currently on the list are: Al-Muhajiroun; British National Party (BNP); English Defence League (EDL); Hizb-ut-Tahir; Muslim Public Affairs Committee; and National Action.

However, this formal and constrained policy statement has inspired similar approaches to those accused of harbouring viewpoints perceived by a critical mass of the student base as abhorrent. These traditionally include homophobia, transphobia, and the expression of principles considered antithetical to progressive thought. However, caught up in this are a number of people who represent moderate critical voices as opposed to those of extremism. An extremely truncated list of people that have faced deplatforming, successfully or otherwise, in the past few years includes:

1. Ivanka Trump, daughter of President Trump, at Wichita State University Tech[8]
2. Feminist icon and author Germaine Greer at Cardiff University[9]
3. Former Home Secretary Amber Rudd at Oxford University[10]
4. Breitbart writer and Gamergate provocateur Milo Yiannopoulos at various institutions[11]
5. Feminist scholar Selina Todd at Exeter College[12]
6. Right-wing author Jordan Peterson at Cambridge University[13]
7. Right-wing author Ben Shapiro at California State University[14]
8. Right-wing author Ann Coulter at UC Berkeley[15]
9. Evolutionary scientist and popular science writer Richard Dawkins at a radio event in Berkeley, California[16]
10. Breitbart editor and advisor to President Trump Steve Bannon at the Booth School of Economics[18]

11. Feminism critic Christina Hoff Sommers at Lewis and Clarke college[17]
12. Former CIA director John Brennan at the University of Pennsylvania[18]
13. Julie Bundel, a radical feminist author, was unlawfully deplatformed by Nottingham City Council over her views on transgender rights[19]
14. Joanne Cherry, MP for the Scottish National Party, had her event at the Edinburgh Fringe cancelled by the Stand Comedy Club in which it was to be held. Upon challenge, they reversed that decision[20].

One might see a trend in this short abridgement. Right-wing perspectives, views from 'the establishment', alt-right provocateurs, and scholars that have been critical of gender theory are all regular targets for deplatforming. In many respects, it is a sign of students being taken seriously by academic decision makers in that the expressed views of the student body can influence the provision of a platform to controversial speakers. However, it is also another sign that universities are fiercely protective of their brands – social media pressure is the primary vector by which these protests are organised.

This speaks to a somewhat concerning trend in the academy for those that are inspired by F. Scott Fitzgerald's assertion that 'First-rate intelligence is the ability to hold two opposed ideas in mind and still retain the ability to function' (Fitzgerald, 1945). Deplatforming essentially removes the opposition from education. It is though inaccurate to say that this represents an assault on freedom since the conflict here is in two fundamentally incompatible **conceptions** of freedom. One is the freedom to express viewpoints, the other is freedom from the 'harm' of being exposed to those viewpoints. In some cases listed above, the speakers might genuinely be considered extremists under PREVENT and associated programmes. In others, particularly those voices that have been deemed 'Trans Exclusionary Radical Feminists', or TERFs[21], it often comes across as more like silencing critical perspectives that are inconvenient or uncomfortable. The Overton window is a concept for describing the parameters within which acceptable discussion can take place in a particular society, and it seems universities are in some cases moving towards a window that is shut more firmly than it is elsewhere and has been in the past.

If one believes that universities should be places where controversial issues are discussed and debated, then unusually it may be the governmental system that comes to the aid of academia. Sometimes perceived as critical of the role of the university in the modern economy, the Conservative government in the United Kingdom has stepped in to create additional room for free speech by issuing guidance and investigating the possibility of strengthened powers for university regulators. Similarly in the US, additional protections for academia are under active discussion. One might cynically note that this has become an issue of increasing importance as these governments have veered ever more to the right themselves – deplatforming after all disproportionately impacts on right-wing voices.

We can conclude from all of this though that monitoring and policing of viewpoints is endemic in the university system, and that it occurs both top down and bottom up. It's a complex issue with no obvious right solutions. And perhaps, given the way that student views on freedom are evolving, the university itself must shift with the times rather than hold strong as a bastion for what is occasionally argued as outdated idealism.

DISCRIMINATION IN THE IVORY TOWER

The emails that have been leaked by Nemesis within our case study reveal a number of interesting new points of information with regard to the postgraduate students thus far most impacted by the university's disciplinary proceedings. Particularly notable is the email discussion between senior members of the university regarding the recruitment of PhD applicants to Blackbriar's team. When Ian McManus describes the four candidates for the two positions, his notes can be summarised thusly:

1. Sharon McAlpine is as qualified as anyone but she'd also improve the university's diversity profile. Also, 'She's very pretty – she'd certainly brighten up that drab office you keep your students in'. Being a mother, the 'risk' of her becoming pregnant again is also raised.

2. Dirk Tumblewood is the Vice Chancellor's nephew and has an impressive list of contacts and qualifications even if his track record for ethical behaviour is questionable. Blackbriar surely wouldn't want to get on the wrong side of the principal.

3. James Duncan is blind but a great coder, and he would be expensive to hire given his accessibility requirements.

4. Stan Templemore is forgettable.

There's a lot to unpack here. There's casual sexism in the way they describe Sharon McAlpine along with a nod to tokenism – that she might be worth picking because she's a woman rather than the best candidate. There's ableism and potentially legally actionable discrimination in the way James Duncan is portrayed, explicitly linking his recruitment to the department's budget. We can also note the unsubtle way in which Dirk Tumblewood is linked to the VC with a 'haha, don't upset the boss' comment that could be hand-waved away as a joke. Stan Templemore, in these circumstances, perhaps gets off lightly. However, it seems unlikely he'll relish the public exposure of his unremarkability to an online audience of meme authors and snarky commentators.

We also see that McManus was right to prep Blackbriar that the VC would not be happy about his nephew being overlooked, especially since the VC is now on record as having stated that he was hoping that the chosen appointments wouldn't work out. His lack of participation in either selection or interviewing though gives him some deniability with regard to any pressure he may, or may not, have exerted. And in any case, he didn't get his way in the end. Given that these two students were suspended though, one could argue a significant conflict of interest is being played out here in the aftermath. Undoubtedly that will be a pillar of the legal case being brought against the university if the lawyer Karan Chandra gets his way.

And yet, while this situation has very poor optics, it's also the case that it's compatible in outcome if not in spirit with the expressed goals of almost any public body in the modern age. Almost every university will have a gender equality and diversity policy that is aimed at normalising the balance in the staff base and student body. Given the discussion in the previous section regarding the no-platforming of gender-critical and right-leaning scholars, it could be argued in this light that the policy directive to create a welcoming and inclusive campus **requires** the prohibition of the expression of certain viewpoints. The conflict here comes from the competing goals of creating a space for free intellectual inquiry and creating a welcoming home for minority groups. It is not possible to be both at the same time regardless of the hopeful perspectives demonstrated by surveys of student views on freedom of speech.

However, what these emails show here is an approach to inclusion that is driven more by spreadsheets and statistics than any genuine desire for integration. 'She's very pretty' is the kind of recruitment criteria we might have expected someone to bring up in an episode of Mad Men rather than within a professional email between managerial colleagues. Still, the inevitable conclusion of efforts to balance genders within the university system is that gender becomes a powerful 'tie breaker' characteristic in recruitment.

But even with gender diversity programmes codified into most universities' working practices, the situation is still hugely uneven. Four countries in the OECD have a greater than 40% proportion of women researchers, and the highest of these is 46% (Abramo et al., 2016). In the UK, women represent 38.3% of researchers, and in Germany it is fewer than a quarter. Even in the Nordic countries, with an enviable reputation for gender equality, it's still the case that men outnumber women two to one. In Australia, women make up a smaller proportion of senior faculty than men, despite making up a greater proportion of junior faculty (Bailey et al., 2016). In Canada, half of all assistant professors are women but only 28% of the professors[22]. In the UK, women earn on average 15.1% less than men in the same position for academic work[23]. In the US, women hold almost half of the tenure-track positions but less than 40% of the tenured posts[24].

What we see from these statistics is something concerning. The problems that exist in recruitment are not sufficient in many cases to keep women from entering academia. That's encouraging. Systemic issues in the university system prevent women entering the higher ranks of university faculties at the rates we should expect. That's alarming. The statistics become even more concerning if we take even the lightest look at intersectionalism where issues of ethnicity, disability, and sexuality dramatically impact on progression (Catalyst Quick Take, 2017). We don't even see a person of colour (PoC) on the short-list for the positions in our case study. Or at least, we can assume we don't given the equality comments in the email chain we can be fairly sure that skin colour would have been brought up in the rundown.

The university faculty system values grantsmanship and research output over teaching, pastoral care, and service (Harvie, 2000; McKenzie et al., 2018). The former are fields that are primarily dominated by male academics. We saw the effect of this early on in Coronavirus lockdown, where women academics were expected to shift over to family care while balancing their professional duties – something we have already addressed. Men simply aren't expected to participate in this kind of work to the same degree and have thus been free to widen the gap of accomplishment (Minello, 2020).

It is likely attitudes like those of McManus have a powerful impact on this. Equality in many fields is a pipeline problem (Makarova et al., 2016) but not exclusively. It's also an issue where systemic misogyny, racism, and homophobia are embedded into the processes and incentives in a way that results in an almost invisible web of complicating factors. Even this is a charitable interpretation and leaves aside the fact that overt displays of discrimination are still a major problem in the academic system (Brunsma et al., 2017; Howe-Walsh & Turnbull, 2016; Madera et al., 2019; Mellifont et al., 2019; Pyke, 2018). The problems don't exist only in the faculty. We've already discussed how students are regularly more critical of women in evaluations and more willing to tolerate eccentricities in male academics (Mengel et al., 2019).

Even if Sharon does come out of this situation with a PhD and her reputation intact, which is unlikely as we have discussed in previous analyses, academia is an institution where the soft impact of implied sexism is considerable. Unfortunately, other industries don't fare much better – in a 2016 analysis it was found that women made up 47% of all employed adults in the US, but held only 25% of computing roles (Ashcraft et al., 2016). Of those women, Asian women made up 5% of the number and black women made up 3%.

For James, the situation is grimmer still. In 2019, only 19.3% of people with disabilities (PwD) were employed in the US compared to 66.3% of those without[25]. As with the gender pay gap, there's also a disability pay gap. It's approximately 15.5% in the UK according to the TUC[26]. PwDs are more likely to lose their jobs and be denied promotions than abled people. Again, we see in McManus a taste of at least one reason – accessibility compensation is costly and the university system only occasionally has the funding to line up internal development with disability inclusion (Heron et al., 2013). Tools, processes, and buildings all have to be modified to ensure accessibility for PwDs. While this is often legally required in many cases, it's an issue of don't ask, don't tell. If nobody with disabilities needs to use your homebrew software setup, you never need to ask yourself hard questions about the usability of the tools upon which you rely. Bespoke tools are common for analysing results, gathering data points, and running research servers. We don't even know how inaccessible they are in general because it's a question we simply have not asked ourselves in the literature.

What we do know from the case study though is that the tools James was required to use were not accessible as a rule and required a number of compromises and workarounds before he could access the data he required. And that in turn creates a difficult question for James to answer: 'Should I rock the boat?'.

Much as with their self-reported adherence to academic freedom, an implicit rule in academia is that it's a safe-space from which to tell truth to power. Those without a disability have the privilege of not worrying about accessibility, at least until the point they do. That places the burden of advocacy on James who is already in a situation where his personal success in the future, difficult under ideal circumstances, is in the hands of the people to whom he would have to complain. He lacks the career stability and professional security that eases the burden of truth-telling. He can choose to make a big thing about the inaccessibility of the tools he has to use. He risks alienating Blackbriar and the colleagues that would need to spend time redeveloping tools for which the source-code may have long been lost (Heron et al., 2013). Or he can tough it out in the hope that he's perceived as someone 'willing to go the extra mile'. If he doesn't keep quiet perhaps when it comes to picking postdoctoral candidates, Blackbriar will go with the one that **didn't** delay the project for weeks or months while the software underwent an accessibility retrofit.

It's here that we see how these systems create circles of sustaining inequality. Inaccessibilities in the system disincentivise PwDs from taking on research jobs. Because few PwDs are in research jobs, software is written by abled people. Abled people, by virtue of not having to worry about it, often don't consider accessibility as a first-tier design goal. And because they develop inaccessible software, it makes it harder for PwDs to do the jobs for which they apply. Systems of discrimination don't need to be intentionally so to be powerfully impactful.

THE AI EMPLOYMENT BOOM

We can actually dig a bit deeper still into the content of these emails. McManus refers offhandedly to an 'AI sift' that took a pool of 236 applicants down to a mere 28 to be considered by human minds. The process of evaluating a CV for a position can take a long time – if it's done comprehensively. Some informal sources suggest that a recruiter might spend as little as six seconds looking at a resume before deciding if someone is suitable for a vacancy; however, these sources almost uniformly refer to unsubstantiated 'studies' without providing supporting evidence. Formal studies tend to put the time-frame somewhere closer to about five minutes, but that is likely a consequence of tools like AI that can immediately reject candidates based on identified incompatibilities.

For example, if a job requires a master's degree, an AI sift can simply remove everyone who only has a bachelor level education. If the prestige of an awarding body is important, the sift can reject anyone who doesn't have a degree from a world-ranked institution. If you need someone with Object-Oriented Programming experience, then if your skills don't list Java, C++, C#, or another appropriate development language then your resume will go into the trash. What you're left with at the end of this process is the set of resumes that are worth spending a little time considering. In this case, perhaps this has saved the recruitment panel anything between twenty and a thousand minutes of work to arrive at the short list. Depending on how many must agree on a CV for it to progress to the next phase, manual consideration might be an institutional cost of several **days** of effort. Or perhaps you can press a button and do it instantly. By Grabthar's hammer… what a savings.

But of course, algorithms encode assumptions and that's as true in an AI sift as it is in anything else. Algorithms reflect the biases of their training sets, and often in nonintuitive ways. Amazon in 2018 scrapped an AI-based machine learning recruitment bot when it discovered the tool had been training on data sets reflecting mostly male resumes. This resulted in the algorithm assuming by default that men were inherently more qualified than women[27]. Companies selling AI hiring tools often note that AI can remove bias from recruiting by eliminating human judgement that could be swayed by race, gender, sociocultural status, and more. Such claims rarely hold up to more than idle scrutiny because in reality simply identifying these traits and controlling for them is an act of biased judgement. It is also tremendously difficult to decouple observable traits of gender and ethnicity without biasing the result (Drage & Mackereth, 2022). We should always be wary when AI seems to be offering broad solutions to broad problems.

Shane in her remarkably fun book *You Look Like a Thing and I Love You* (Shane, 2019) outlines the 'four signs of AI doom', which are the conditions under which AI will dramatically under-perform at a set goal or indeed will turn out to make things worse.

1. The problem is too hard.
2. The problem is not what we thought it was.
3. There are sneaky shortcuts an AI could exploit.
4. The AI is learning from flawed data.

And indeed, all four of these come into play when it comes to doing an AI sift on resumes. Unless training data is well scrubbed, balanced, impervious to gaming, and can capture the softer implication of meaning contained in the deeper connections on a resume... well, the sift just isn't going to be very good. An acquaintance of the authors has worked at a high level in a complex field for two decades, and still found herself being rejected from jobs below her skill level because they mandated an 'upper second' bachelor degree and she had a lower second. The years of skill development and professional expertise didn't matter. She fell below that threshold, and thus was sifted out.

Let's assume though that **this** AI sift – the one employed in our case study – has beaten the odds and is actually doing exactly what it's supposed to. It is identifying the best candidates, from good training data, in a well-understood domain. Then we see our old friend coming in – 'gaming the system'. CVs are amenable to keyword stuffing – just make sure you put the right words in the right places. Unless very clever, algorithms find it difficult to separate human readable text from the actual binary bits of the data. Including the job advert itself in an invisible white font as part of the background of your resume **might** be enough to get you past the filter. Leaving aside these dubiously ethical practices, it still means that recruitment is more about how you optimise your keywords as opposed to how well you would do the job. Smart candidates will be employing the same process to their resume as someone might modify a website to better match a search engine's expectations. There are even AI tools out there that are used to build resumes, and these are optimised for the requirements and preferences of AI sifts. In such circumstances, recruitment becomes computers talking to computers, rather than people trying to ascertain how each would fit into the professional lives of the other.

We have some obvious candidates listed here for the job at the University of Dunglen, and they demonstrate a clear match for the job. Who **didn't** make the cut though, and what reasons are there for their exclusion? Would these software decisions stand up to scrutiny in the event someone makes a formal complaint? The lack of transparency in AI that we have already discussed comes into play here – companies rarely give transparent access to their data sets and can rarely communicate the decision points in any decision. Treleaven et al. (2019) note that:

> Resume-sifting algorithms exhibit unethical, discriminatory, and illegal behaviour; crime-sentencing algorithms are unable to justify their decisions; and autonomous vehicles' predictive analytics software will make life and death decisions.

Sachoulidou (2023) has argued that big data has cut away at the presumption of innocence that is the basis of justice in most of the developed world. It may very well be that, encoding biases as they do, algorithms in sensitive areas like recruitment, sentencing, and medicine might actually be in clear violation of long-standing equality legislation. Modern discourse centres on the role of generative AI in accelerating an existential crisis for human relevance. The crisis began long ago – we're only really just catching up to how much of our lives are automated in unknowable ways.

NOTES

1 https://web.archive.org/web/20220524223241/
2 https://www.timeshighereducation.com/news/vice-principal-quits-protest-failure-punish-colleagues
3 https://www.kentonline.co.uk/canterbury/news/christ-church-university-ex-vice-a56351/
4 https://www.theguardian.com/technology/2017/may/13/accidental-hero-finds-kill-switch-to-stop-spread-of-ransomware-cyber-attack
5 https://en.wikipedia.org/wiki/Listofnewsmediaphonehackingscandalvictims
6 http://www.legislation.gov.uk/ukpga/2015/6/contents/enacted
7 http://www.sussex.ac.uk/prevent/
8 https://www.newyorker.com/news/news-desk/ivanka-trump-and-charles-koch-fuel-a-cancel-culture-clash-at-wichita-state
9 https://www.theguardian.com/books/2015/nov/18/transgender-activists-protest-germaine-greer-lecture-cardiff-university
10 https://www.theguardian.com/politics/2020/mar/06/amber-rudd-hits-out-at-rude-oxford-students-after-talk-cancelled
11 https://www.vox.com/policy-and-politics/2018/12/5/18125507/milo-yiannopoulos-debt-no-platform
12 https://www.bbc.com/news/uk-england-oxfordshire-51737206
13 https://www.theguardian.com/education/2019/mar/20/cambridge-university-rescinds-jordan-peterson-invitation
14 https://abc7.com/ben-shapiro-csula-escorted-protest/1219358/
15 https://www.campussafetymagazine.com/university/uc-berkeley-ann-coulter-appearance-security/
16 https://www.bbc.com/news/world-us-canada-40710165
17 https://www.insidehighered.com/news/2018/03/06/students-interrupt-several-portions-speech-christina-hoff-sommers
18 https://www.thedp.com/article/2016/04/protests-shut-down-cia-director-john-brennan-talk
19 https://www.nottinghampost.com/news/nottingham-news/nottingham-city-council-issues-apology-7679208

20 https://www.bbc.co.uk/news/uk-scotland-edinburgh-east-fife-65451979
21 https://www.newstatesman.com/politics/2017/09/what-terf-how-internet-buzzword-became-mainstream-slur
22 https://www150.statcan.gc.ca/n1/daily-quotidien/191125/dq191125b-eng.htm
23 https://www.ucu.org.uk/genderpay
24 https://nces.ed.gov/ipeds/use-the-data
25 https://www.bls.gov/news.release/pdf/disabl.pdf
26 https://www.tuc.org.uk/research-analysis/reports/disability-employment-and-pay-gaps-2019
27 https://www.ml.cmu.edu/news/news-archive/2016-2020/2018/october/amazon-scraps-secret-artificial-intelligence-rehtml

REFERENCES

Abdenour, J. (2017). Digital gumshoes: Investigative journalists' use of social media in television news reporting. *Digital Journalism, 5*(4), 472–492.

Abramo, G., D'Angelo, C. A., & Rosati, F. (2016). Gender bias in academic recruitment. *Scientometrics, 106*(1), 119–141.

Abrams, D., Wetherell, M., Cochrane, S., Hogg, M. A., & Turner, J. C. (1990). Knowing what to think by knowing who you are: Self-categorization and the nature of norm formation, conformity and group polarization. *British Journal of Social Psychology, 29*(2), 97–119.

Ashcraft, C., McLain, B., & Eger, E. (2016). *Women in tech: The facts.* National Center for Women & Technology (NCWIT).

Bailey, J., Peetz, D., Strachan, G., Whitehouse, G., & Broadbent, K. (2016). Academic pay loadings and gender in Australian universities. *Journal of Industrial Relations, 58*(5), 647–668.

Beetham, H., Collier, A., Czerniewicz, L., Lamb, B., Li, Y., Ross, J., Scott, A.-M., & Wilson, A. (2022). Surveillance practices, risks and responses in the post pandemic university. *Digital Culture and Education, 14*(1), 16–37.

Bösch, C., Erb, B., Kargl, F., Kopp, H., & Pfattheicher, S. (2016). Tales from the dark side: Privacy dark strategies and privacy dark patterns. *Proceedings on Privacy Enhancing Technologies, 2016*(4), 237–254.

Branch, S., Ramsay, S., & Barker, M. (2013). Workplace bullying, mobbing and general harassment: A review. *International Journal of Management Reviews, 15*(3), 280–299.

Brunsma, D. L., Embrick, D. G., & Shin, J. H. (2017). Graduate students of color: Race, racism, and mentoring in the white waters of academia. *Sociology of Race and Ethnicity, 3*(1), 1–13.

Bryan, H. (2017). Developing the political citizen: How teachers are navigating the statutory demands of the counter-terrorism and security act 205 and the prevent duty. *Education, Citizenship and Social Justice, 12*(3), 213–226.

Catalyst Quick Take. (2017). *Women in academia.*

Christensen, T., Gornitzka, Å., & Ramirez, F. O. (2019). Reputation management, social embeddedness, and rationalization of universities. In *Universities as agencies* (pp. 3–39). Springer.

Constantinides, E., & Zinck Stagno, M. C. (2011). Potential of the social media as instruments of higher education marketing: A segmentation study. *Journal of Marketing for Higher Education, 21*(1), 7–24.

Davies, M. (2018). Educational background and access to legal academia. *Legal Studies, 38*(1), 120–146.

Donald, W. E., Baruch, Y., & Ashleigh, M. (2019). The undergraduate self-perception of employability: Human capital, careers advice, and career ownership. *Studies in Higher Education, 44*(4), 599–614.

Donelan, H. (2016). Social media for professional development and networking opportunities in academia. *Journal of Further and Higher Education, 40*(5), 706–729.

Drage, E., & Mackereth, K. (2022). Does AI debias recruitment? Race, gender, and AI's "eradication of difference". *Philosophy & Technology, 35*(4), 89.

Fanghanel, J., & John, P. (2015). Editors' conclusion higher education and the market: thoughts, themes, threads. In *Dimensions of marketisation in higher education* (pp. 233–240). Routledge.

Fitzgerald, F. S. (1945). *The crack-up*, ed. E. Wilson. New Directions.

Friedman, S., Savage, M., McArthur, D., & Hecht, K. (2019). *Elites in the UK: Pulling away? Social mobility, geographic mobility and elite occupations*. International Inequalities Institute.

Gearing, A. (2014). Investigative journalism in a socially networked world. *Pacific Journalism Review, 20*(1), 61–75.

Gearon, L. F., & Parsons, S. (2019). Research ethics in the securitised university. *Journal of Academic Ethics, 17*, 73–93.

Gilmore, J. (2017). Teaching terrorism: The impact of the counter-terrorism and security act 2015 on academic freedom. *The Law Teacher, 51*(4), 515–524.

Giroux, H. A. (2006). Academic freedom under fire: The case for critical pedagogy. *College Literature, 33*(4), 1–42.

Graham, A., Powell, M. A., & Truscott, J. (2016). Facilitating student well-being: Relationships do matter. *Educational Research, 58*(4), 366–383.

Greer, S., & Bell, L. (2018). Counter-terrorist law in British universities: A review of the" prevent" debate. *Public Law, 2018*(January), 84–104.

Grugulis, I., & Stoyanova, D. (2012). Social capital and networks in film and TV: Jobs for the boys? *Organization Studies, 33*(10), 1311–1331.

Hargittai, E., & Marwick, A. (2016). "What can I really do?" Explaining the privacy paradox with online apathy. *International Journal of Communication, 10*, 21.

Harvie, D. (2000). Alienation, class and enclosure in UK universities. *Capital & Class, 24*(2), 103–132.

Heineman, K. J. (1992). *Campus wars: The peace movement at American state universities in the Vietnam era*. NYU Press.

Heron, M., Hanson, V. L., & Ricketts, I. (2013). Open source and accessibility: Advantages and limitations. *Journal of Interaction Science, 1*(1), 1–10.

Howe-Walsh, L., & Turnbull, S. (2016). Barriers to women leaders in academia: Tales from science and technology. *Studies in Higher Education, 41*(3), 415–428.

Keller, T. (2019). Finding news sources. In *Television news* (pp. 271–302). Routledge.

Lieberwitz, R. L. (2002). The corporatization of the university: Distance learning at the cost of academic freedom. *Boston University Public Interest Law Journal, 12*, 73.

Luo, L., Wang, Y., & Han, L. (2013). Marketing via social media: A case study. *Library Hi Tech, 31*(3), 455–466.

Madera, J. M., Hebl, M. R., Dial, H., Martin, R., & Valian, V. (2019). Raising doubt in letters of recommendation for academia: Gender differences and their impact. *Journal of Business and Psychology, 34*(3), 287–303.

Makarova, E., Aeschlimann, B., & Herzog, W. (2016). Why is the pipeline leaking? Experiences of young women in stem vocational education and training and their adjustment strategies. *Empirical Research in Vocational Education and Training, 8*(1), 2.

Masur, P. K. (2019). New media environments and their threats. In *Situational privacy and self-disclosure* (pp. 13–31). Springer.

McKenzie, A., Griggs, L., Snell, R., & Meyers, G. D. (2018). The myth of the teaching/research nexus. *Legal Education Review, 28*(1), 1–20.

McLelland, L. B. (2019). *The campus speech wars: The problem of freedom in higher education* (Doctoral dissertation). University of Alabama Libraries.

McNeill, T. (2012). "Don't affect the share price": Social media policy in higher education as reputation management. *Research in Learning Technology, 20*, 152–162.

Mellifont, D., Smith-Merry, J., Dickinson, H., Llewellyn, G., Clifton, S., Ragen, J., Raffaele, M., & Williamson, P. (2019). The ableism elephant in the academy: A study examining academia as informed by Australian scholars with lived experience. *Disability & Society, 34*(7–8), 1180–1199.

Mengel, F., Sauermann, J., & Zölitz, U. (2019). Gender bias in teaching evaluations. *Journal of the European Economic Association, 17*(2), 535–566.

Milkman, R. (2017). A new political generation: Millennials and the post-2008 wave of protest. *American Sociological Review, 82*(1), 1–31.

Miller, D., Xu, X., & Mehrotra, V. (2015). When is human capital a valuable resource? The performance effects of Ivy league selection among celebrated CEOs. *Strategic Management Journal, 36*(6), 930–944.

Minello, A. (2020). The pandemic and the female academic. *Nature Worldview*. https://doi.org/10.1038/d41586-020-01135-9.

O'Neill, K. F. (1983). The ambush interview: A false light invasion of privacy. *Case Western Reserve Law Review, 34*, 72.

Pasquini, L. A., & Evangelopoulos, N. (2017). Sociotechnical stewardship in higher education: A field study of social media policy documents. *Journal of Computing in Higher Education, 29*(2), 218–239.

Pomerantz, J., Hank, C., & Sugimoto, C. R. (2015). The state of social media policies in higher education. *PLoS One, 10*(5), e0127485.

Pyke, K. D. (2018). Institutional betrayal: Inequity, discrimination, bullying, and retaliation in academia. *Sociological Perspectives*, *61*(1), 5–13.

Rapoport, A. (1989). Prisoner's dilemma. In J. Eatwell, M. Milgate, & P. Newman (Eds.), *Game theory* (pp. 199–204). Palgrave Macmillan

Ringmar, E. (2007). *A blogger's manifesto: Free speech and censorship in the age of the Internet*. Anthem Press.

Rutter, R., Roper, S., & Lettice, F. (2016). Social media interaction, the university brand and recruitment performance. *Journal of Business Research*, *69*(8), 3096–3104.

Sachoulidou, A. (2023). Going beyond the "common suspects": To be presumed innocent in the era of algorithms, big data and artificial intelligence. *Artificial Intelligence and Law*, 1–54. https://doi.org/10.1007/s10506-023-09347-w

Shane, J. (2019). *You look like a thing and I love you*. Hachette.

Spiller, K., Awan, I., & Whiting, A. (2018). 'What does terrorism look like?': University lecturers' interpretations of their Prevent duties and tackling extremism in UK universities. *Critical Studies on Terrorism*, *11*(1), 130–150.

Treleaven, P., Barnett, J., & Koshiyama, A. (2019). Algorithms: Law and regulation. *Computer*, *52*(2), 32–40.

Wai, J., & Perina, K. (2018). Expertise in journalism: Factors shaping a cognitive and culturally elite profession. *Journal of Expertise*, *1*(1), 57–78.

Wilson, J. K. (2015). *Patriotic correctness: Academic freedom and its enemies*. Routledge.

Zimba, A., & Mulenga, M. (2018). A dive into the deep: Demystifying WannaCry crypto ransomware network attacks via digital forensics. *International Journal on Information Technologies and Security*, *10*(2), 57–68.

CHAPTER 12

Witch-Hunts at the University

IT Crackdown Causes Criticisms

NEWSPAPER ARTICLE

An exclusive report by Jack McKracken for the Dunglen Chronicle

Students and staff alike today are in an uproar after what has been classed as a systematic witch-hunt by the University of Dunglen's technical support team. The damaging leaks of last week have resulted in what one student on Blether called a 'massively disproportionate drum-head trial' of anyone found violating the Acceptable Usage Policy (AUP) of the institution.

Stan Templemore is a Master's student at the university, doing a postgraduate qualification in data mining. He spoke to the Chronicle today. 'I know a lot of people who have been suspended, or warned, for violating the AUP, even for trivial, stupid stuff like streaming FlixWatch movies or watching YewChoob. They get threatened that they'll be suspended and their names will go up in a newspaper as a warning to others. It's totally unfair'.

Terry Holmes, head of Information Technology at the University, replied to our request for interview with an extract from the university's newly revised policy. It contains the following clauses:

The University network may not be used directly or indirectly by a user

DOI: 10.1201/9781003426172-12

for the creation, transmission, or receiving of:

1. Obscene or offensive images, data, or other material.
2. Material with the intent to harass, upset, or offend a third party.
3. Material which promotes discrimination on the grounds of race, sex, religion, belief, disability, age, sexual orientation, or political affiliation.
4. Material which advocates or supports the performing of illegal or unlawful acts or activities.
5. Material which is libellous or in some other way defamatory.
6. Material which infringes the intellectual property rights of the university or third-party intellectual property owners.
7. Material which consumes undue amounts of shared university resources.
8. Material which violates the university's security protocols or provides advice on how this can be done.
9. Material which is digitally manipulated to convey a false impression.

Clauses seven, eight, and nine of the AUP were newly instituted in the last week after a decision to unilaterally alter the policy was taken by the university governors.

'They blocked off a pile of ports in the system', says Templemore. 'Anything that wasn't used for the world wide web, basically. Even if you wanted that, you had to go through a proxy – that meant they had full control over what sites you could visit and could track which IPs went where. That might be okay for most people, but computing is a big department here and has a big cohort of students. We need access to servers and protocols that others don't, and the IT department have refused to open the ports we need'.

Jane Dymock, a second-year computing student studying Digital Ethnography and Software Engineering, said, 'Some of us live in halls and we only have access to the university networks. The new policy means that we couldn't play games, watch videos or listen to music. It's only to be expected that people would find a way around it. We're computing students, after all, and the university gets a bunch of money from the government to provide the resources we need for our education. Imagine if this policy had come about during the pandemic, when we were all stuck isolating!'

Terry Holmes has responded to criticism of the new policy, saying, 'The resources provided by the university are made available so that the lawful, day-to-day activity of the university can be pursued. It is true that some relatively benign services have been caught up in these restrictions, and we will assess these on a case-by-case basis. When they are reviewed and we decide they pose no significant threat to our networks, we'll re-enable access.

We're not looking to shut down all non-productive use of the systems, just make sure that we have more control over who is accessing our resources and how'.

He continues, 'After extensive investigation of our network access logs, we find no evidence that anyone has fraudulently gained access to our servers. As such, we must conclude that the emails leaked by the individual calling themselves Nemesis are entirely invented. However, in our internal security audit we found several routes through which a dedicated hacker could gain access to restricted systems. While we deal with those, it is necessary to take a harder line with our AUP than has previously been the case.

This is only a temporary restriction, but it is also necessary that we deal harshly with those students seeking to circumvent the protections we install. We can't be seen to be soft on violations of security'.

This is part of a wide platform of changes at the university. In addition to the updated AUP, Dunglen's plagiarism policy has been revised to prohibit any and all usage of large language models in student assessment, along with similar blanket bans on techniques such as AI-enabled text-to-image generation, automated code generation, virtual agents, or machine learning. Some universities have put these decisions in the hands of individual academics, but this new policy requires special permission from the university Senate for exceptions. These will be considered on a case-by-case basis by a newly instated Committee for the Responsible Use of AI in Education. Alongside these developments, the university has introduced formal policies on how it will employ 'Ethical AI' in academic and research circumstances.

The change to the plagiarism policy is widely seen as a response to the worry that essay mills will give way to more sophisticated generation of coherent assignments through generative text and associated methods. 'It's already possible to get an algorithm to debug a C++ function', said one anonymous lecturer, 'It's really making it difficult for us to assess student work on the basis of what the student – rather than an AI – can do'.

Much of our understanding regarding the current upset about the university's access policies is based on a series of emails leaked by the mysterious individual known as Nemesis. Nemesis claims to have taken these directly from the email account of Mr. Holmes, which is something that the director of IT denies strenuously although he has not denied the authenticity of the emails themselves. The first leaked email relates to the imposition of the new clauses in the AUP:

```
To: Terry.Holmes@dunglen.
ac.uk
```

From: i.p.freely24122@hotmail.com
Subject: Now u c me Hey Terry!

Saw your group email today! Megalolz! Way to win the hearts and minds buddy.

See you, Nemesis.

< Dear all
There has been a serious breach of the university's Acceptable Usage Policy regarding IT usage. Private, confidential data from this university may have been leaked to the press. This is not only against university policy, it is against the law and we will be seeking to prosecute anyone who has had any involvement in this incident. If you have any knowledge of this, please contact me immediately.

We are willing to be lenient if the culprit comes to us before we need to get the Police Caledonia Cybercrime Investigation division to investigate.

Further to this, the university will now adopt a zero tolerance policy to any misuse of its information technology infrastructure. We already assist intellectual property holders in identifying copyright infringement done through the university network, and block many of the ports used by standard file sharing applications.

From 9am tomorrow, we will be blocking all ports outside the university except for those used by the world wide web. For access to the web, it will be necessary to use the university's proxy server. All other ports will be restricted. See the newly revised AUP for the clauses which concern this.

Regards,
Terry. >

The email above was sent to all students. The above was followed up by another two alleged exchanges with the hacker:

To: Terry.Holmes@dunglen.ac.uk
From: now.u.don7@gmail.com
Subject: Way to go!

Haha!

You suspended those guys for using StreamSong? That's a legal service, dude!

I know they had to bypass your dumb proxy to do it,

but if you knew anything about computer security they wouldn't have been able to!

Hearts and minds man, hearts and minds.

Your buddy, Nemesis.

< Dear student

Your usage of non-sanctioned services (e.g. FlixWatch, StreamSong) constitutes a breach of the university's Acceptable Usage Policy. You took advantage of security holes in our web proxy in the process, and this is a disciplinary and legal offence. Your registration with this university has now been deactivated pending review by the university's governors. Your names will be made available to the police, and handed over to the local paper for them to publish in their new name and shame section.

Regards,
Terry. >

The name and shame section indicated in the above email was an idea proposed by Mr. Holmes to the Dunglen Chronicle, but one that we did not wish to participate with because of its distasteful implications. The final alleged exchange between Nemesis and Mr. Holmes is reproduced below:

To: Terry.Holmes@dunglen.ac.uk
From: in.ur.grill@outlook.com
Subject: Drumhead Trial

Woah

You're taking this way too far, Tezza. You can't blackmail someone with being suspended from the university if they don't shop their friends to you. Nobody knows who I am – you'll never get to me through them. I'm way too smart for you.

Your buddy,
Nemesis.

< Dear student

The university governors have permitted me to reenable your account provided you cooperate fully with the university's investigation into illegal hacking of our servers. You've already shown yourself to be capable of circumventing certain countermeasures, and this makes you a suspect in our inquiry. Please come to my office at 9am tomorrow. I will ask you who else you know in your social circles who would be capable of gaining access to our internal systems. Full cooperation with this will result in the reinstatement of your student enrolment and ceasment of any criminal charges we brought to bear. If you choose not to cooperate, your suspension will stand and you will be removed from your course of study.

Consider the consequences,
Terry >

'These computing resources are provided using in many cases public money', says Mr. Holmes in response to the allegations of draconian mishandling of the situation. 'Student funding covers their use of the systems, but only for university work. The AUP contains a clause which says we can change the agreement unilaterally at any time without notice if it is necessary to protect the university infrastructure. In this case it is'.

Stan Templemore disagrees. 'I can't do most of my work within the university now', he says. 'The services I need to use are unavailable to me. I can't even get access to my home email through the Wifi. I'm an international student so I'm paying 'full economic fees' for my education. Something I wouldn't have been doing as recently as a few years ago, but that's Brexit for you. I'm now in a position where I'm paying for access to resources that aren't suitable for what I'm doing. Anything that touches the outside world I need to do at home. I'm already behind where I wanted to be in my career – I had to take a suspension from my studies due to mental health issues in the pandemic. Now I'm being prevented from even catching up to where I want to be'.

Templemore continued. 'The LLM restrictions in particular are unconscionable. I have colleagues who are working with AI in their thesis projects, and they're working within the ethical AI guidelines put out by the university. Even so, the way the plagiarism policy is written means that they need to completely revise the work they have been doing and making it much less relevant and much less interesting. All because the people drafting these policies haven't been involved in a real research project for decades and don't know what proportionality means'.

'Some power users may be disproportionately impacted by the new policy', concedes Mr. Holmes. 'We'll identify the areas where that kind of thing happens and look to work with those users to resolve their problems in a security conscious manner. We're not trying to make the network unusable, just hardened against attack by people who want to do our services and staff harm'.

Professor Blackbriar last night was unavailable for comment.

THE UNIVERSITY-HOSPITALITY COMPLEX

What Coronavirus has revealed to many people, in many cases for the first time, is the extent simple student presence on a campus has on the financial health of the university system. The abrupt shift to remote teaching (Bozkurt & Sharma, 2020) was a complex story of mixed successes and failures (Hodges et al., 2020). It has also had numerous universities, particularly in the United States, calling out for financial assistance as their hospitality offerings ceased to generate sustainable revenue. Battles over the refunding of tuition fees are likely to begin in earnest in the years following the crisis. The cracks in the residential offerings – which include critical infrastructure resources such as library and internet access – are perhaps more easily exploitable. None of this is settled, and it's likely we will see a number of changes in the relationship between universities, their students, and the economic systems within which both work. It's an age in which we are suddenly hyper-aware of the possibility of global pandemic and some have argued that nothing will be the same again.

With this in mind, the issue we're going to address in this chapter is the way in which the university has approached its information breach problems through policy setting. One anonymous student (again, always anonymous) has referred to a 'drum-head trial' of anyone found violating an AUP that has been recently modified to prohibit a range of activities that had previously been tacitly approved.

In fairness, the AUP of the University of Dunglen is not out of line with those of real-world institutions. Many of us have signed policies of equivalent wording when undertaking employment or buying an internet package. However, the issue here is one of interpretation. 'Material which consumes undue amounts of shared university resources' is one that carries with it a lot of leeway. What is an 'undue amount'? Who decides? And what recourse do students have in the event that they are found in violation of this policy? One must bear in mind that at most institutions students are paying fees which include access to computers, network resources, and storage – to what extent can the university deny them and to what extent is it appropriate that everyone be impacted by the actions of a single individual? Given the student tuition fees are paid as a single bundle without bolt-ins or equivalent, is it fair that a university denies access to one part of a service when opting out was never part of the equation?

The new policy seems to be in response to the leaked emails from Nemesis, which Terry Holmes has claimed are completely invented. However, the service shutdown has impacted on students watching

FlixWatch (our fictional equivalent of Netflix), playing games, and streaming music. One might argue that these are not necessary activities in an educational establishment, but increasingly universities are tied into the hospitality industry. Dorms, refectories, night-clubs, and more are all part of the student provision. Here, students in university dorms – linked into the university network are now being denied recreation at home on the basis of what's happening in their universities. Coronavirus has shown how much universities depend on the income from these sectors and the battle over refunds is likely to be bruising.[1,2,3] Students in these circumstances have no alternatives – they cannot simply shift internet access to another provider. The auxiliary fees levied by universities can be considerable – they are hard-baked into the budgetary assumptions under which institutions function. Residential services are usually the largest slice of these auxiliary funding streams, and the fact that they also permit managed control over resources is a valuable benefit. On top of this, universities offer campus bookstores, event hosting, on-campus hotel services, parking, and vending machines – all of these tie into the operating budget of the institution.

At Harvard in 2019, board and lodging fees counted for almost 4% of revenue. Amherst College makes up 9% of its operating budget from housing-related income. In 2018, Smith College in America collected over 40 million dollars in residential and dining fees, and that corresponds to approximately 16.5% of the total revenue for the institution. These are substantive revenue streams – the University of Massachusetts was expecting to lose around 70m USD to refunds as a result of Covid-19-related issues.[4]

In our case study Holmes acknowledges some 'power users' will be disproportionately impacted by the policy and that services will be resumed gradually. This is phrased as a kind of hardening of the network to protect it against intrusion. Under emphasised in the reporting is that Holmes himself said there had been no illegal access of university services which raises the question of the necessity of the restrictions at all.

And again, this is not uncommon in real universities. University networks, already populated by mischievous students and staff, are common targets for external hackers and internal pranks. A university represents an incredibly complex information ecosystem and much of it is likely to be legally protected or commercially sensitive. Student records, research data, emails, and staff personal directories are difficult to fully protect given the openness of the academic system. Interfaces between public and private services are often points of trouble, and academics must collaborate both nationally and internationally on projects that may involve a baffling array of legal complication and technical compromises.

Consider a scenario where a UK academic is using a piece of research software they have written for a Linux server they run. It cross-references student attainment versus social media activity – all ethically above board – and then compiles the analysis of these data points into an interrogable source that can be used for academic publication. And now imagine that they are collaborating with three other academic partners – one in China, one in Sweden, and one in the USA. And then imagine they all need to share equivalent data with each other and have it examined both individually and then in aggregate to explore the link between social media presence and grades across national boundaries.

It's not an uncommon structure for a project, but it's one sufficient to make any security professional, IP lawyer, or compliance specialist weep at their desk.

For one thing, given that it's an academic writing the software – how secure did they make it? Does it leave ports open, intentionally or otherwise? Does it expose an API? Is it properly sanitising input? Is it using the latest versions of all its libraries given that most exploits are rooted in pre-existing software packages?[5] What software stack does it use and how does it make the connections between the various layers? What tool-chain is used for development and what provision is made for cleaning the outputs?

What of the server? Is this appropriately protected and running a secure OS? Who else is using it and what permissions do they have? Have all the software package **they** have used been secured? How reliable is the architecture? Consider how much of the internet fell apart when popular software package Log4j failed.

These are not idle concerns. One of the authors of this book almost lost a private server to hackers because he had set up a default WordPress install for the use of someone else. They had temporarily set the username and password to the same thing, expecting that to then be changed by the new admin – not secure, but still normal practice. The other user neglected to change the password for several months. The result was that malicious actors used that username and password combination to gain access to the admin menu of the second WordPress install, which gave access to the server, which gave access to every piece of software running on that system. There was no legally compromising data on the server, but it did have implications for the research project that the server was supporting (Heron et al., 2018a, 2018b).

Assuming the software and server are secure – does the data comply with the requirements of applicable legislation? What of data gathered in China and processed in the UK? What of data gathered in Sweden and processed in China? Is it all compliant with EU directives? Is it ideologically compliant with Chinese requirements?

How is the data being sent, and what are the ethical and legal complications there? Is it being sent directly between servers? How is it being sent, and are those servers employing encryption? Could the data be intercepted? How commercially sensitive is it and how compromising would it be for participating research subjects? If it's being transferred by Dropbox or Google Drive or an equivalent – is that even legal in the circumstances?

Given these kinds of escalating complexities, and the inability of any one individual to keep on top of them, organisations are increasingly moving to a lockdown model of security. Even Google, in 2023, announced a pilot scheme to create an airgap between employees and the wider internet.[6] Companies sometimes institute proxies that all traffic must go through. They require whitelisting of connections to non-standard ports or enable then only on networks disconnected from the main one. They create or commission their own offerings of tools such as Doodle, Dropbox, Media Wiki, and more and then attempt to force staff into using them. The power users that cannot function with the tools provided are assured that their issues are being looked into. Difficulties in collaborating between incompatible software ecosystems will be marked as 'being addressed'. The extent to which free functioning of knowledge work can function hand in hand with the legal responsibilities of a corporation is a constant source of friction. Sometimes it is the end-users that bear the heaviest burden as the debate is worked out. In the case of our Scandal, what we are seeing is a form of collective punishment expressed through legalese as the university tries to regain control of a situation that has long ago left it scrabbling for an effective response.

UPDATED EXPECTATIONS

The changes we can see in the AUP and the new policies introduced by the university reflect a response to an educational environment that has become almost unrecognisable of late. We have already discussed many of the changes that the Coronavirus brought about – some of them temporary, such as an abrupt and often unruly shift to online teaching (Chhetri, 2020) – but also in terms of shifting attitudes to our work and life balance. That shift was intensely disruptive, and the early signs of its impact are concerning (Reimers, 2022). However, the authors of this book would argue that this pales in comparison to the changes to come with the sudden emergence of AI in its post-2022 form. AI is actually performing well in a number of areas now, and it seems like that's going to be an existential problem for us as a species.

The long-standing joke in computer science as a discipline has been that artificial intelligence is 'ten years away' from living up to its promise, and it's been 'ten years away' since the 1950s. It's never been true – while AI has excelled at certain specialist, narrow tasks such as playing chess and producing mathematical proofs, there was always a huge gap between the promises of the technology and the reality. After decades of effective AI being ten years away, it turns out the people saying it ten years ago were finally correct. AI is now good at a lot of things, and some of those things are fundamental to what we consider the baseline skills of an educated populace. AIs are now good at writing, for example. They're good at expressing ideas and building coherent arguments. The output from ChatGPT and the many alternative systems write convincingly and with clarity, even if as we've seen already they don't necessary have the ability, yet, to write **accurately**. Of course, it's wrong to suggest that this is a purely technological accomplishment – in the background is an underpaid and exploited army of people responsible for cleaning data and feeding it into the algorithms (Williams et al., 2022).

AI is now good at writing code and even fixing broken code that is submitted to it. It can even do it in incredibly obscure contexts. For years, the first author of this book was the primary developer on a Multiuser Dungeon (MUD) called Epitaph. This ran on a driver (FluffOS) upon which sat a Mudlib (Epiphany). The driver provided a language called LPC, which is poorly documented, and the specifics of it vary from mudlib to mudlib. As part of the training material associated with the Epitaph and the MUD from which that itself was derived (Discworld MUD), Michael wrote a full textbook on LPC within the Epiphany mudlib (Heron, 2010). While the generated code wasn't **quite** correct – it was correct **enough** to be a massive aid to productivity. Generated code may need some cleaning up, some syntax shifts, some changes to the inherits, and so on but an area in a MUD may comprise hundreds of rooms, and now GPT can output those in a matter of seconds. Detailed area work, with rooms crafted by human developers, means new areas in a game can take months, or years, to develop. With AI, it could be done in a few days. Especially if you could, for example, tell the AI to 'Add a command that lets someone write on the blackboard' and it would alter the code to allow it to do exactly that.

With a combination of prompt engineering, human editing, and mudlib design, it's entirely possible for a MUD – again, a massively obscure and niche form of gaming – to be constructed at speed and at scale using nothing more specialised than the baseline GPT available to the general public. If someone wanted to develop their own LLM trained entirely on MUD area code, one has to assume something much more effective and less human intensive could be generated.

How do you examine a student's ability to write, or to code, or to create user interfaces? How do you assess them in any number of previously human responsibilities when an AI can do it quicker, more cleanly, with no errors, and with no obvious sign of AI intervention? And if that's impossible, do we even really need to be teaching this stuff at all? Generative AI is likely to require a fundamental reassessment of how we **do** assessment. It may be some time before we quiet the reverberations of this sudden explosion of AI competence – if we even can.

The University of Dunglen is not alone in perhaps overreacting to the worry – governments throughout the world are scrambling to work out legislation to govern AI, and sometimes even to work out with whom the jurisdiction for such legislation belongs. We cannot comment too deeply on this – at the time of writing, most of this legislation is still in early draft form – but suffice to say there is much that needs to be addressed. Some of these concerns are old. Some are desperately new. LLMs are trained on huge data sets, and those data sets often include copyrighted material. It is possible, for example, that my own written material on LPC is the basis for some (or even perhaps a lot) of GPT's fluency with the language. I say it is possible, because the general public are not permitted access to this training data to see if it is non-infringing, and that's a problem that needs to be addressed. As soon as something is fixed into a tangible form, copyright protections apply. A Large Language Model using your website for its training is already violating your intellectual property. Perhaps you have an open licence, such as a Creative Commons derivative. Good luck getting an attribution to your work in the output of an infringing generative AI.

We have seen this debate reach perhaps its sharpest tone in relation to those AI systems that generate imagery, as it is not simply a case of reproducing images but also **styles**. Styles and techniques cannot be copyrighted, but social prohibitions against this form of 'moral plagiarism' are quite strong in the world of art. Dall-E, MidJourney, and other platforms on the other hand will quite happily create images in almost any style with no complaints or hesitation. And, importantly, with no acknowledgement of source or of the intellectual debts incurred. Artists have cause to be worried – a piece of Midjourney-generated art won a fine arts competition last year, and that outcome resulted in an outpouring of artistic anger in response.[7]

However, in this we have to accept that it is simply not possible to put the genie back into the bottle. The word 'disruptive' is overused in modern technological discourse, but it is exceptionally apt when it comes to the future of this specific technology. Just as society is now divided into two eras – post- and pre-pandemic – we are now divided into post- and pre-AI.

Bostrom (2019) put forward an interesting way of thinking about technological development. Imagine every time a new technology is invented, it's pulled in ball form out of an urn filled with invisible and infinite possibilities. Some of the balls come out white – they are unquestionably beneficial to society. Medicines and vaccines, agricultural innovations that solve famine, that sort of thing. Most of the balls are grey, which is to say they have mixed blessings. They have some good features, and some bad features, but the extremes of both can be absorbed by society without existential threat. However, lurking in that urn are also black balls – technologies that will inevitably destroy humanity as a direct and causative consequence of their invention. We have been exceptionally fortunate as a society that all the world-ending technologies that have been invented are costly, complicated, and governable. Imagine where we would be if the ingredients for a nuclear bomb could be sourced from the average home garage. The difficulty and detectability of refining weaponised chemical compounds though is the preventative measure that ensures that a home-grown nihilist can't destroy a city, and a network of such nihilists cannot destroy a world.

Generative AI though is a massively disruptive technology, arriving in a society that is not yet adapted to its presence. It is also, comparatively, exceptionally cheap and easy to replicate. Everyday citizens probably don't have enough in their bank account to train something of the scale of GPT3, but the cost is in the very low millions as opposed to the billions. Some estimates suggest that GPT3 cost around $25m to train the first time. It's not peanuts, but it's not equivalent to the estimated $25bn (adjusted for inflation) associated with the Manhattan project. As computer resources continue to get cheaper, that cost will go down. As the cost goes down, more private citizens will have access to the resources needed to build their own specialist systems. People will find new ways to use these techniques to disrupt ever greater portions of the human experience – the AI aggregator 'There's an AI for that' already has thousands of entries for specialist AI tools. We've focused on text and images in this book, but there are hundreds of problem domains in which AI techniques are starting to dominate.

Some downplay the risk of AI, pointing out that even at its best AI will fail to match the skill and creativity of a talented human. The authors of this book believe that to be true. However, we also believe that's less reassuring than people think. The threshold at which AI obsoletes a human is not that it exceeds our competence, but rather that it passes the threshold of being 'good enough'. At that point, the value comparison becomes 'Excellent for lots of money' or 'Good enough for nothing'. In most cases, the people in control of the purse strings will likely opt for the latter option.

Our most pressing priority now, as a species, is in finding a way to coexist with artificial intelligence. In that frame, the University of Dunglen is perhaps gripping too hard on the past. Banning AI outright is like trying to extinguish the stars in the sky. It won't work and you'll just look foolish trying. As Stan Templemore points out, in order to understand something you need to study it. Banning all use of generative output is not going to solve the problem. It's just going to kick a can a very short way down a very long road.

AN ETHICAL AI POLICY

Related to this, we see here that the university has also implemented an ethical AI policy although we are not permitted to see it. As such, we can use it as a kind of 'theorycrafting' exercise to discuss what such a policy might mean, and what it ought to target. For that, we can start to bring together many of the themes we have discussed already. An ethical AI policy should try to ensure several things:

1. An absence of bias in training materials
2. Transparency of data sets and algorithms
3. Explain-ability and auditability of decisions and outputs
4. Respect for privacy

5. Respect for intellectual property
6. Respect for societal norms

One of the first attempts to codify rules of behaviour for machines is to be found in Isaac Asimov's 'Three Laws of Robotics'. It's striking to see the difference between these predictive clauses and the priorities facing policy makers in the next few years:

1. A robot may not injure a human being or, through inaction, allow a human being to come to harm.
2. A robot must obey orders given to it by human beings except where such orders would conflict with the First Law.
3. A robot must protect its own existence as long as such protection does not conflict with the First or Second Law.

We have not reached the point at which we need to worry so much about autonomy in machines – AI is a fragmented field of incompatible techniques that do not cohere into a general replica of human intelligence. When algorithms make decisions, they do not reason about them the same way a human does. The concept of harm and injury is something that is not governed by intention, but by gaps in training data or overconfidence in the application of rudimentary algorithms to complex use-cases. That doesn't diminish the associated harm but rather re-centres the responsibility for it onto those that build and use AI systems.

An absence of bias would ensure that algorithms reflected society's priorities rather than hidden quirks in a data set. Pursing this as a goal would also force us to confront those biases that are hidden and those that are encoded second-hand in other information.

Consider an algorithm designed to assess whether someone should be given a business loan. It is trained on a data set made up of people who repaid previous loans and those that defaulted. It doesn't capture any data about names, or ethnicities, or gender. What it does capture is addresses, professions, educational levels, work experience, and so on. It seems to lack a racial bias, but only if we neglect to consider that there is often ethnic clustering around particular geographical areas. It seems to be gender neutral, but a look at the underlying sex-disaggregated data would reveal some professions are dominated by women rather than men and vice versa. Bias comes in many places, in many forms, and controlling for it requires deep engagement with data and why we need to record it and why we need to use it.

LLMs are trained on vast data sets, and the composition of those data sets is often inaccessible to end-users. That's a problem. It's also a problem when algorithms responsible for large-scale decisions about humans lack transparency. We've seen many examples of jurisdictions in the United States using algorithms for supporting decision-making associated with when prisoners should be granted parole and how convicted criminals should be sentenced (Bonnefon et al., 2020; Villasenor & Foggo, 2020). When one wishes to appeal these decisions, there's no way to object to the way in which an algorithm arrived at its conclusions because the data sets and the processes are proprietary. It is a black box that a desire for human mercy cannot unlock.

We are also increasingly seeing a world in which AI is so complex it cannot even be understood by the people that develop the systems. Instead, we have AIs responsible for designing other AIs. The link between what goes in and what comes out is often baffling, and many systems offer no way to break down the decision-making into a human-readable format. We expect humans to be able to explain the logic of their decisions, and if we expect less of machines, we'll find ourselves in a world governed by unknowable systems of implacable severity.

The power of AI, and the speed at which it can work, creates its own classification of issues for privacy. It's now possible for surveillance technology to assess facial composition in real time, cross-reference it with other databases, and form a sophisticated picture of the activities of millions of people. While these systems are not perfect, that is perhaps a point **against** the emergence of a digital panopticon. Consider China's social credit system (Liang et al., 2018) and the impact a classification error may have on the life of an otherwise exemplary citizen.

Finally, if we are to have any hope that humanity will continue to produce works of art and science and culture, AI needs to be respectful of intellectual property and of the general philosophy of collaborative culture sharing. Otherwise we will see our world descend into a hauntology as described by Mark Fisher (2014) – where we lose the ability to meaningfully invent the future because artificial intelligence anchors us permanently in the past. With nobody creating anything new, generative AI becomes the only source of new culture. That in itself can only ever be a remix of what has come before.

Is this what the University of Dunglen's policy is going to cover? We have no idea... but we do know where AI is heading, and we know what we need to do to harness it. We can only hope that those responsible for its development are equally thoughtful of its broader long-term implications.

THE DEATH OF BELIEVABILITY

One clause in the new AUP we haven't devoted much attention to yet is the last one. The AUP prohibits:

> Material which is digitally manipulated to convey a false impression

We've already talked a bit about how wide some of these clauses are, and this is perhaps the one that is actually the most important and most far-reaching. On the face of it, it seems reasonable – the university does not want its systems being used to transmit intentional misinformation. However, a lot of content is going to be caught by this. At one end, one might argue an image post-processed in Adobe Photoshop has been 'digitally manipulated' to 'convey a false impression'. And indeed, in some cases one might even say this kind of routine operation transgresses into the realm of the unethical. Use of Photoshop and Instagram filters has been linked to everything from creating a distorted sense of beauty standards (Rajanala et al., 2018) to literal white-washing of dark skin in corporate imagery. In 2009, Microsoft was caught replacing a Black office worker for a white worker in a promotional campaign in Poland. L'Oreal stopped using Photoshop to exaggerate the effect of their mascara in promotional literature after a widespread backlash. There are numerous incidents of universities in America pasting imagery of Black students into promotional material to make their campus seem more diverse. In 2023, an image of the UK's Prime Minister Rishi Sunak poorly pouring a pint was digitally changed to make it appear like flawless pull. Sometimes the manipulation is less blatant – for example, brightening the hues across an entire photograph will tend to lighten skin tones. Where do we draw the line between everyday, acceptable digital manipulation and outright abuse?

One suspects though that it's not this specific problem that the university is trying to shut down – what is far more likely in the current climate is that they are worried about deep fakes.

For those that haven't encountered them, deep fakes are a form of media in which characteristics of one digital file are used to manipulate another. For example – changing the text of a word document so it is spoken with the voice of a real person. You might find an audio file in which someone says 'I love Vladmir Putin', and another file with Donald Trump giving a speech. With judicious use of some clever software, you can produce a convincing third file in which Donald Trump says 'I love Vladimir Putin' despite never having said such a thing. That's… alarming. If anyone can be made to say anything, in an environment where fact-checking of online material has become its own battleground (Tandoc, 2019), then we may be in the age where you simply cannot believe anything you see or hear.

This becomes especially alarming when used in the context of video. It is possible to take the face of one person and have that digitally mapped on top of the face of someone else in a different video. If you want footage of Boris Johnson saying something outrageously offensive, you just need to find a video of someone vaguely similar in appearance saying it and transplant Johnson's face onto the original. Or you can just Google for a while, because you'll surely find an authentic video somewhere that doesn't need any alteration at all. Again, this isn't an idle concern – it's already happening. A deep fake video of Ukrainian president Volodymyr Zelensky asking his soldiers to lay down arms was distributed on social media. Turkey's president Recap Erdoğan, during a closer-than-anticipated election, directed his supporters to a video (again, fake) showing his opponent receiving an endorsement from the Kurdistan Workers Party (PKK). PKK is classified as a terrorist organisation in Turkey, the UK, and the US. Joe Biden verbally attacked transgendered women in another deep faked video.

You can see the opportunity for mischief here, and as the algorithms get better the harder it becomes to tell fact from fiction. However, the technology – in addition to the role it can play in the creation of disinformation – has basically invented a whole new category of sex crime inflicted upon women in the public eye. Specifically, using public footage spliced on to pornography to create fake pornographic videos. Technically speaking this is also a sex crime from which men can suffer but suffice to say – like many of the things we have discussed in this book there is a very clear gender bias in the demographic makeup of the victims. We've already seen in a previous article that Sharon McAlpine has been the victim of this.

There are now numerous sites on the internet dedicated to this form of 'Deepfake Pornography'. No matter the explicit act you want to see, you can easily find a famous woman performing it. If you have a specific combination in mind, well – the technology is available to you at home. If you have access to the necessary raw material, the technology does not only work on the famous. If you want to see a friend naked, you can make it happen – no agreement is needed from them for you to press a button on a piece of software. That said, there are indeed legitimate and consensual uses of the techniques. Cersei Lannister's naked 'walk of shame' in Game of Thrones was partially done through the application of CGI, and Peter Cushing's appearance in the movie Rogue One was only made possible through techniques that are now made more affordable through the framework of deep fakes.

There is certainly an appetite out there for video and images of celebrity sex. In 2014, hundreds of private photos of celebrities were leaked to the internet. Some of these were simply candid personal photography. Others incorporated nudity and sexually suggestive poses. These were clearly intended for the enjoyment of a specific person or persons. As soon as the images hit the internet, they became impossible to retract. Multiple torrents containing the collection remain easily available, and several sites have curated this collection and integrated it with other leaked photos from other sources. Combined with celebrity sex tapes, there is a considerable archive out there of nonconsensual pornography. In this, it is (presumably) not the imagery itself that lacks consent, but rather that consent was intended to apply to a very limited audience. Upon entering the public record, the conversion of the imagery to freely available pornography is where the consent broke down.

Response to these leaks is mixed, with some saying, 'If you didn't want people to see you naked, you shouldn't have taken the pictures'. Victim blaming, unfortunately, is a common response to issues of sexual crime directed at women. Deepfake pornography however invalidates this line of argumentation because no famous actor consented to having their personal imagery placed atop the sexual performance of someone else.

It is hard to argue that this is not a use to which deep fakes were intended to be put. Sometimes similar effects can be obtained through algorithms designed for different purposes. Most respectable generative AI platforms are issuing strict restrictions on how the tools may be used to support the generation of NSFW text and imagery. Adobe's Photoshop licence bans it outright, as do the terms of service associated with ChatGPT, Bing Chat, Dall-E, Midjourney, and Stable Diffusion. This is necessary, because while these platforms were never designed with sexual content in mind, the algorithms do not differentiate the sexual from the non-sexual unless explicitly designed to do so. Stable Diffusion in particular is a good case study in the complexity of interacting systems of ethics – it is an open-source platform, which means that its source-code can be forked and the protections against generating sexual imagery can be removed. In some versions of the tool, the filter can be disabled by changing a single line of code. The prosocial benefits of open source here come into collision with the need to protect the rights of individuals. In case you were wondering – yes, these tools can be used to edit photographs and yes they can be used to replace clothed people with naked people.

One suspects that the potential for harm associated with deep fakes is something the university wants to control as much as possible and to ensure it's clear that they do not endorse or support it. Imagine a speech from the principal in which he says, 'Yes, Blackbriar was right. We screwed him over. And we'll do it again!'. A combination of algorithms that control lip syncing, face transplants, and audio transformation could be employed to create additional levels of misinformation in an already complicated situation. The university can't stop it happening, but it certainly put rules in place that discourage people from doing it with the university's own resources.

NOTES

1. https://www.business-standard.com/article/international/covid-19-crisis-angry-undergrads-suing-colleges-for-billions
2. https://www.insidehighered.com/news/2020/03/13/students-may-want-room-and-board-back-after-coronavirus-closures-refunds-would-take
3. https://www.forbes.com/sites/richardvedder/2018/06/14/why-are-universities-in-the-housing-business/
4. https://www.bizjournals.com/boston/news/2020/03/27/umass-to-lose-70m-in-revenue-to-student-refunds.html
5. https://info.veracode.com/report-state-of-software-security-open-source-edition.html
6. https://arstechnica.com/gadgets/2023/07/to-defeat-hackers-google-wants-employees-to-work-without-internet-access/
7. https://www.creativebloq.com/news/ai-art-wins-competition

REFERENCES

Bonnefon, J.-F., Shariff, A., & Rahwan, I. (2020). The moral psychology of AI and the ethical opt-out problem. In S. M. Liao (Ed.), *Ethics of artificial intelligence* (pp. 109–126.). Oxford University Press.

Bostrom, N. (2019). The vulnerable world hypothesis. *Global Policy, 10*(4), 455–476.

Bozkurt, A., & Sharma, R. C. (2020). Emergency remote teaching in a time of global crisis due to coronavirus pandemic. *Asian Journal of Distance Education, 15*(1), i–vi.

Chhetri, C. (2020). "I lost track of things" student experiences of remote learning in the COVID-19 pandemic. In *Proceedings of the 21st Annual Conference on Information Technology Education (SIGITE '20)* (pp. 314–319). Association for Computing Machinery, New York, NY. https://doi.org/10.1145/3368308.3415413

Fisher, M. (2014). *Ghosts of my life: Writings on depression, hauntology and lost futures*. John Hunt Publishing.

Heron, M. J. (2010). *The epitaph survival guide*. Imaginary Realities.

Heron, M. J., Belford, P. H., Reid, H., & Crabb, M. (2018a). Eighteen months of meeple like us: An exploration into the state of board game accessibility. *The Computer Games Journal, 7*(2), 75–95.

Heron, M. J., Belford, P. H., Reid, H., & Crabb, M. (2018b). Meeple centred design: A heuristic toolkit for evaluating the accessibility of tabletop games. *The Computer Games Journal, 7*(2), 97–114.

Hodges, C., Moore, S., Lockee, B., Trust, T., & Bond, A. (2020). The difference between emergency remote teaching and online learning. *Educause Review, 27*. https://er.educause.edu/articles/2020/3/the-difference-between-emergency-remote-teaching-and-online-learning

Liang, F., Das, V., Kostyuk, N., & Hussain, M. M. (2018). Constructing a data-driven society: China's social credit system as a state surveillance infrastructure. *Policy & Internet, 10*(4), 415–453.

Rajanala, S., Maymone, M. B., & Vashi, N. A. (2018). Selfies—Living in the era of filtered photographs. *JAMA Facial Plastic Surgery, 20*(6), 443–444.

Reimers, F. M. (2022). Learning from a pandemic. The impact of COVID-19 on education around the world. In F. M. Reimers (Ed.), *Primary and Secondary Education During Covid-19: Disruptions to Educational Opportunity During a Pandemic* (pp. 1–37). Springer.

Tandoc, E. C. Jr (2019). The facts of fake news: A research review. *Sociology Compass, 13*(9), e12724.

Villasenor, J., & Foggo, V. (2020). *Artificial intelligence, due process and criminal sentencing. Michigan State Law Review*, 295. https://heinonline.org/HOL/LandingPage?handle=hein.journals/mslr2020&div=11&id=&page=

Williams, A., Miceli, M., & Gebru, T. (2022). The exploited labor behind artificial intelligence. *Noema Magazine, 13*.

CHAPTER 13

Hacker Is Postgraduate Student

NEWSPAPER ARTICLE

An exclusive report by Jack McKracken for the Dunglen Chronicle

The mysterious hacker known as 'Nemesis' was today revealed as postgraduate student James Duncan, who has been illegally accessing university systems for the duration of his suspension. Duncan was identified as a result of the university's internal and external surveillance of staff and students. The university itself released a statement:

The University of Dunglen, like most universities, employs a sophisticated network of monitoring to ensure that all system accesses are compliant with our IT usage policy. We also maintain an ongoing monitoring presence of staff and students with regard to social media. We look to ensure that all mentions of the institution are compliant with the Social Media Policy to which all users agree upon registering their affiliation.

Terry Holmes, head of Information Technology at the university, explained how Duncan was identified: 'Every time someone accesses one of our systems, they leave a trace. When they access remotely, they leave the details of which computer they used to make the connection. Our hacker used a "remote proxy" which means that they routed their connection through another computer elsewhere in the world, and so we couldn't identify their exact location even with the

DOI: 10.1201/9781003426172-13

intrusion detection software we were employing. Really we were stuck, until we put out a public statement saying that the hacking hadn't occurred and that the data being released was fraudulent. The hacker then posted a picture to a social media account that we monitor, showing their computer screen as they accessed our systems'.

The indicated statement was released yesterday by the university and said the following: 'After extensive investigation of our network access logs, we find no evidence that anyone has fraudulently gained access to our servers. As such, we must conclude that the emails leaked by the individual calling themselves Nemesis are entirely invented'.

'The hacker couldn't let that stand', said Holmes, 'And just had to prove us wrong.

Unfortunately, he forgot to strip the meta-data from the image he posted to Facebook, so we got his GPS co-ordinates from that. We passed them on to the police, and the police picked him up late last night'.

James Duncan has been embroiled in the growing catastrophe of crisis at the University of Dunglen, having been suspended for his role in a supposed case of academic fraud. His identification as the hacker Nemesis comes as a surprise to his colleague, Mrs. Sharon McAlpine.

'James was the most technically adept of us', she said, 'But I had no idea he knew how to do stuff like that!'

James Duncan spoke to us this morning after being released on bail and explained how he had gained access to the system. 'It wasn't complicated really – for people working with a lot of computer science, most of the people in our department don't know a lot about computers. They're engineers and mathematicians and chemists, not programmers or sysadmins. So they'd come to me when they needed something done. I knew a lot of the passwords and passkeys through the system, and I also knew when people had left their systems open for exploitation. Humans are always the weakest chain in any security regime. Some of them I didn't even need to know in advance – people use the same passwords for everything, and they're not very careful in choosing them. Ask someone where they grew up, the name of their childhood pet, and their mother's maiden name – now you've got access to their security answers'.

Police seized all of Duncan's computer equipment and personal files as part of the arrest, and he is currently looking at prosecution under Sections 1, 2, and 3 of the Computer Misuse Act 1990, as amended by the Serious Crime Act in 2015. As part of his invasion into the Dunglen computer network, he made modifications to systems and logs to hide his tracks, which counts as 'unauthorised modification' under the legislation. Many of his activities may also fall under the general category of identity theft.

We asked Detective Inspector Cameron, of Police Caledonia, who spoke to us today regarding the incident. 'Not only was Mr. Duncan responsible for leaking confidential and private material, it seems like he was inventing emails to support his own case and weaken that of the university. Given the situation we would request that all parties treat email conversations leaked through the press as unreliable because we cannot confirm which of these are true and which are not'.

All parties in the scandal have seized on the arrest as an opportunity to distance themselves from damaging leaks. The University of Dunglen has released a statement insisting that the evidence of faked emails entirely exonerates the senior management of the university of wrong-doing. Professor Blackbriar's lawyers have insisted that email correspondence pertaining to the research data is accurate, although those relating to discussions between himself and Mrs. Sharon McAlpine were invented. Karan Chandra was today unavailable for comment.

An insider at the university told us, 'There's a palpable sense of relief. Some of the emails were definitely accurate – completely true. Others weren't. Some were even generated by AI to some kind of agenda. I can't say which are which, but it's very easy to fake something like this if you don't need to square it up against server logs and such. Coupled to this, the best place to hide a lie is between two truths – I think the McAlpine emails are the only ones we know were definitely faked, but it's so hard to tell with anonymous informants. The fact that there's so much confusion and distrust though means that any evidence coming from leaks or hacks can be hand-waved away'.

The revelation of the extent of the university's surveillance programme has come as a shock to many within the beleaguered institution. 'They're scanning my PeerNet and Blether accounts for mentions of the university? What on earth for?', asked one first-year Geography student. 'What I put on there is my business'.

When asked whether she had read the Acceptable Usage Policy for the university or the Social Media Usage Policy to which she agreed upon university registration, she said, 'No, but it's not like I get to decide if it's okay. If I don't sign it, I don't get to register. If I don't get to register, I don't get to do my degree. It's totally disproportionate'.

Brand management for organisations has become increasingly important over the past 15 years, with social media profiles requiring constant curation. 'A bad comment on Blether has a habit of escalating out of control', said Terry Holmes. 'We have people who need to make sure that complaints are heard, and that staff members aren't being indiscreet with our corporate information. It's just like having security staff on University grounds – we need to make sure our investments, physical and otherwise, are protected. Nobody ever forgets on the internet, and rebranding an institution of our size is largely untenable as a response to scandal. Our best bet is to proactively protect our name'.

Professor Blackbriar remains unavailable for comment.

THE LAWS AND ETHICS OF COMPUTER HACKING

We discussed in Chapter 3 about a kind of scaffolding that takes us from 'morality' through 'ethics' into 'laws'. Specifically, laws can be viewed as a framework of externally enforced ethical codes. Whereas betraying our professional codes of ethics might only carry the sting of being ostracised, that's not the same when it comes to the law. The state, in whatever form its enforcement takes, chooses the extent to which we are to be punished for infractions. With the activities of Nemesis, we enter a different kind of consideration than we have discussed so far. That is to say the extent to which his activities were compatible with law. And, unfortunately, it's not looking good for him. But also, again, it's not as simple as 'he broke the law' or 'he didn't break the law'.

One thing that defines the legal framework around computer crime is that it is usually outdated. The time-frame in which legislation, case law, and precedent operate is sedate and stately. It can take years for a complex case to work its way through the various magisterial stages of the legal process. The European Court of Human Justice outlines in Article 6 of its convention that 'everyone is entitled to a fair and public hearing within a reasonable time' (Mahoney, 2004). The International Covenant on Civil and Political Rights states as part of Article 9 'Anyone arrested or detained on a criminal charge shall be brought promptly before a judge or other officer authorized by law to exercise judicial power and shall be entitled to trial within a reasonable time' (Macken, 2005). However, that word 'reasonable' is carrying a lot of weight on its shoulders in both of these statements.

The way that legal systems are constructed is often in two parts. One is what we might think of as 'setting laws', and in modern democracies there is usually an intentional lag built into this process so as to ensure that complex issues receive due consideration and contemplation. At least, in theory. In the United Kingdom for example, a bill goes through several parliamentary stages including a first reading, second reading and debate, analysis by committee, a report stage, a third reading, and then it is voted on by both the House of Commons and then the House of Lords before receiving Royal Assent. Many of these stages also include room for amendment, rejection, and re-evaluation. There are numerous other ways in which proposed legislation can be delayed and adjusted within the process of refinement. Or, of course, new laws can be rejected entirely, forcing a rethink that may require the expenditure of additional time or political capital. Most modern democracies have processes like this that ensure a single dominant party, or individual, can't change the law in a kneejerk fashion. It's not always the case that the slow consideration of cause and effect is given time to mature, of course, but that's the intent of the system.

In most cases, legislation, unless it is forced through, requires **time**.

But the thing about computer science is – it **doesn't** require time in the same way. The pace of advancement in technological fields is rapid – far beyond the ability of governments to regulate it. Tech often moves too fast for humans to keep up with it in daily life. Alvin Toffler coined the phrase 'future shock' to describe the phenomenon of 'too much change in too short a period of time' (Toffler, 1970), In 2012, David Karpf introduced the concept of 'Internet Time', which is the rate at which online connectivity is reshaping society (Karpf, 2012). The pace of change identified by these individuals is only increasing (Butler, 2016; Leadbeater, 2020). Moore's Law is often trotted out in these kinds of discussions, referring to the idea that available computer power doubles every 18 months or so. The statement can be interpreted in a number of ways, but the traditional metric is 'number of transistors that can be packed into in a given area'. There are other ways in which the law manifests, and perhaps a more robust way of thinking about it is 'What does £100 buy you in terms of processing power?'. Much is made of the fact that Moore's Law is slowing down (Eeckhout, 2017; Flamm, 2018) and losing its predictive power. Much less is made of the fact that the doubling of computer power can take many different forms whether through efficiencies in manufacturing, design, or energy consumption. Processors may be getting faster at a slower rate, but computing power at scale has never been so achievable – or so affordable.

Our legislative frameworks do not advance nearly so quickly, and so we are left with situations where existing laws – such as the Computer Misuse Act (Macewan et al., 2008) – are sadly out of date and no longer reflect the contemporary digital landscape (Montasari et al., 2022). The Computer Misuse Act is over 30 years old and may be forgiven for the creaks and cricks it carries in its bones. However, even the relatively spritely General Data Protection Regulation (GDPR), introduced in 2018, is already struggling to keep apace with technological development (Kutylowski et al., 2020). The introduction of generative AI into the equation has further complicated things. In many ways, LLMs and the implications thereof have acted like a sniper's bullet aimed at the heart of computing legislation.

Crimes such as those of James Duncan must be viewed in light of laws that were written for scenarios that are no longer fully operable. In this, we might harken back to the justification for this book itself, way back in the first chapter… *The Case of the Killer Robot* (Epstein, 1994) requires so much explanation about its context that it has lost all its power as an aid to understanding. Computer legislation suffers a similar context drift with regard to computer crime.

Before we get too deep into this, it is perhaps appropriate at this point to outline some important disclaimers.

First of all, this isn't a text on law or legal interpretation. As such, an analysis of applicable legal texts to this case study is outside the scope of the work and the domain of author expertise. Secondly, the case study is explicitly set in the UK and as such it is UK legislation that is pertinent – while many countries have comparable laws, they all differ in important respects. And thirdly, nobody should take such analysis as we provide as anything approximating 'legal advice'. To paraphrase the vernacular of the internet, WANAL (We Am Not A Lawyer).

We all though face the complications of legality in what we do, and so we still need to ground ourselves in at least some interpretation.

The Computer Misuse Act of 1990 outlines three offences with regard to computer access. The first is that it restricts 'unauthorized access to computer material'. The second is that it specifically outlines 'unauthorized access with intent to commit or facilitate commission of further offences'. It has a third part which looks at unauthorised access with the intention of doing criminal damage to a computer system itself.

The act has been amended, most recently in 2015, to include two additional offences. These cover 'unauthorized acts causing, of creating risk of, serious damage' and 'making, supplying or obtaining articles for use in offences under all four sections above'.

What then did James Duncan do within our case study that could be considered a violation of the act?

The first thing hinges on 'unauthorized access', and here things become murky. Most Acceptable Usage Policies (AUP) tend to include clauses on not giving people your password, but as we discussed previously it's hard to consider the way an AUP is worded to be an exemplar in eliciting 'informed consent'. 'I knew a lot of the passwords and passkeys through the system', James said. His colleagues would come to him when they needed help. The extent to which he was authorised to use their accounts, and to what extent they were authorised to permit him access, is an important factor here. In many cases, he didn't hack their passwords. He used passwords he had been given. That means that he is perhaps safe under section 5 of the act – no keyloggers, rootkits, or brute-force password crackers seem to have been employed. And the extent to which he caused 'serious damage' in the sense we'd understand it is questionable.

It is clear though that this isn't the full story – he notes that humans are always the weakest part of any security system, and our human fallibility makes gaining genuinely unauthorised access easier. There's a whole fascinating discipline known as 'social engineering' which is what you might get if you crossbred hacking and con-artistry. James seems at least conversant with some of its basic principles. The important part of the act here is that it makes no distinction about **how** unauthorised access was obtained. Section one of the act certainly seems to apply, by James' own admission.

Section two of the act is again, murky. What are the offences that James was committing? To address this, we need to take a little diversion into the World of Wikileaks and what it was James actually did.

We might be charitable here and ascribe to James a noble motive – he was seeking to draw attention to the real-world complexity behind the public reporting of John Blackbriar. In other words, we might see him as a kind of whistleblower – someone who provides information on people or organisations that are regarded as engaging in immoral or illegal activity. There are legal protections available to someone who does this, although they vary in strength, applicability, and intensity. Within the United Kingdom – the legal jurisdiction in which we are most interested here – there exists the Public Interest Disclosure Act 1998 which grants professional protection to those that break confidentiality clauses to inform on dubious behaviour of which they are aware. A whistleblower is, at least in theory, protected from being treated unfairly or losing their job because of their activities.

However, whistleblowing legislation rarely includes explicit protections from punishment if other laws are broken in the process of obtaining or disclosing information. The expectation is very much that people should work within the systems framework that the legal system has constructed, even if that framework prevents gaining access to the information needed to properly disclose important facts. In short, James may have done the right thing (or at least, some could argue that he did) but in the eyes of the law that doesn't justify or exonerate his activities.

It's in this capacity that we can understand the situation related to other high-profile whistleblowers. Edward Snowden was an intelligence consultant who, in 2013, leaked information from the United States' National Security Agency (NSA). This information was highly classified, identifying American involvement in a wide range of questionable activities. These included membership in the Five Eyes Intelligence Alliance

(Walsh & Miller, 2016); an email scanning system called PRISM (Berghel, 2013); the indexing of massive amounts of metadata as part of the Boundless Informant programme (Masco, 2017); hacking of China (Landau, 2013); and much more. Snowden also revealed highly embarrassing information regarding the role of the UK's Government Communication Headquarters (GCHQ) in surveilling the diplomats of 'friendly' governments. The leaks also provided information as to the extent to which GCHQ had hacked into international internet infrastructure. It is safe to say that this leak was profoundly embarrassing to a number of intelligence organisations across the world. However, in the gathering and releasing of this information Snowden broke the law – classified material is not the same thing as confidential material. The charge levied against him was breaking the Espionage Act (Scheuerman, 2014), and in response he fled first to Hong Kong and then to Russia where, as of the time of writing, he remains as a full citizen.

There is little doubt that Snowden revealed many illegal and unconstitutional activities, but in breaking the law he lost any whistleblower protection that may have applied. Should he return home to the United States, he would undoubtedly have to face prosecution for his crimes. The ends, the legal system would suggest, do not justify the means. That said, some have argued that the Espionage Act needs amending to enshrine protections for people like Snowden (Barnes, 2014; Zeman, 2015). This is a common thread as we have seen – that legislation cannot keep pace with how technology changes. To quote ourselves for a moment (Heron, 2016):

> Twenty years ago, Manning would have had to remove confidential State department memos with a convoy of wheelbarrows. Now, gigabytes of data can be transferred using a thumb drive no bigger than a finger.

Thus, we see here the tensions for whistleblowers who may see themselves to be doing good while operating within systems designed to make revelation extremely dangerous. The site Wikileaks, for example, is a prominent repository of classified and confidential information. Its public profile is often contradictory. It sometimes positions itself as merely an interface between journalists and open-access data. Other times it presents itself as a 'public intelligence agency'. Nonetheless, it has had a prominent role in serving as a shield for those that wish to release classified information in relative safety. Data leaks it has been involved with include media footage showing the murder of civilians in Baghdad airstrikes (Allan &

Anden-Papadopoulos, 2010); the Afghanistan War Logs (O'Loughlin et al., 2010); and millions of classified diplomatic cables (Lynch, 2013). During the 2016 presidential campaign, they also revealed leaked documents from the Hilary Clinton campaign – the latter incident being notable for how tightly woven it is into the allegations of collusion between Donald Trump and the Russian state intelligence apparatus (Ioffe, 2017). Chelsea Manning was a source of important documents for WikiLeaks in 2010 and was imprisoned for seven years (after a presidential commutation of the earlier 35-year imprisonment) in Fort Leavenworth. Snowden escaped the US legal system. Manning fell foul of it. Julian Assange, the controversial founder of WikiLeaks, spent seven years under the diplomatic protection of the Ecuadorian embassy in London before his asylum was withdrawn in 2019. Since then, and as of the time of writing, he has been confined in HM Prison Belmarsh, a maximum security prison in London, awaiting extradition to the United States.

That these disclosures served the public good is debated – WikiLeaks in particular has long been criticised for improper disclosure (Benkler, 2011; Hindman & Thomas, 2014) and endangerment (Brevini, 2017). The inevitable political context in which the releases must exist is also an ever-present talking point, with approval for these acts of whistleblowing often breaking along ideological lines. What is clear though is that illegally releasing information 'in the public interest' is no protection for whistleblowers, and whatever motives we might ascribe to James Duncan he's going to find them scant protection against the ire of the judiciary. Under section 2 of the Computer Misuse Act, he's probably got a lot to worry about.

As to the third section of the act – doing criminal damage – well, in trying to cover his tracks he might just have made things worse. Removing mentions to himself in routine system logs certainly would count as an 'unauthorized modification' and realistically may fall into the category of enabling the committing of other offences. Covering ones trail, it could reasonably be argued, is a necessary prelude to further intrusion.

In other words, James probably needs to engage the services of a lawyer who knows their way around this legislation. And he probably needs to do it **soon.**

Critics of the Computer Misuse Act argue that it is too broad and doesn't properly differentiate between those that hack for noble reasons (known as White Hat hackers in the parlance of information security) and those that hack for malicious reasons (the Black Hat hackers). The very framing here as 'unauthorized access' was put in place to make an attempt to future-proof the legislation. Unfortunately, its vagueness has meant that there have been few individuals that have been successfully prosecuted or jailed under the law (Montasari et al., 2016). A generation of children have grown up knowing that leaving their phones unattended while they are logged into social media is a recipe for having their accounts vandalised by their friends. If only those friends realised that they were engaging in an activity that, under law, can carry with it a 12-month custodial sentence under section 1 of the act. This from a piece of legislation that can't even address many modern cyber morbidities within its framing (Worthy & Fanning, 2007). For some, this legislation is most distinctive for how little it does to create a prosecutory framework for the most important technologies that govern our lives.

INFORMED USER CONSENT

What is notable here is perhaps not so much the crime itself, but the investigation around the crime. James fell victim to his own hubris perhaps, believing his computer skills rendered him largely invulnerable to capture. However, we often underestimate how much data we are each producing on a day-by-day basis... and how little knowledge we have about what we generate. Current estimates suggest that human society creates around 120 zettabytes of data a day. That number is made up from a lot of different sources. That's 500 hours of new YouTube footage **every minute**. It's 183,000 blog posts per hour. Ten billion Facebook status updates in a day. In 2023, that's 1.7 megabytes of data, per person, **per second**. To go along with all our intentional and considered contributions, most of what we produce is what we call 'metadata' – or data that helps us make sense of other data.

Let's look at a pertinent example of that. When we take a digital photograph, our phones and cameras try to be helpful. They encode a lot of additional data about the image to help software packages make sense of it. Image metadata includes information on aperture. It contains info on dots per inch. It contains shutter speeds and focal depth. It contains data format information and sizing. That's all important and necessary for when you start working with your photos in Lightroom. The metadata though contains other things, such as the GPS co-ordinates of where you were when you took the photo. The set of data captured is documented by a standard known as EXIF (Exchangeable Image File Format), and it can be comprehensive. If you post a photo without clearing out your EXIF data, there's a good chance other people can access that metadata too. Like James, you may find you're revealing more than you hoped. Or, in the cases where EXIF data is modified... less than others may hope.

After all, this is how the fugitive John McAfee's location was leaked to the world (Madhisetty et al., 2019). A picture taken by a Vice journalist revealed that the people in the photo were in Guatemala. McAffee later claimed that he had spoofed the EXIF data, but that's a claim that doesn't much hold up to scrutiny.

Many of us don't think a lot about the information we're leaking out simply by being a living person in an age of computers. Metadata is everywhere. Sometimes it seems innocuous, such as a list of authors in a word processing document. Sometimes that document will contain metadata regarding tracked changes, which... well, you may not want just anyone being able to read your in-progress draft notes. Sometimes it's genuinely harmless, like the list of fonts installed on your computer or how much memory is in your device. Sometimes though when that information is cross-linked with other information it can do frightening things ... like uniquely identifying your computer from nothing more than the metadata provided when you access a website (Eckersley, 2010). This unique identifier is something known as a digital fingerprint. If you can fingerprint someone, you can track them. Did you stay off of Facebook because you didn't want it knowing everything about you? Bad news... they worked it out anyway. You've got a 'shadow account' and there's not a thing you can do about it (Sarigol et al., 2014; Tufekci, 2018). The Facebook pixel is one of the tools used to do this – a 1x1 invisible image that site operators can include on a site to hook users into the vast machines that do behavioural analysis across social media.

Facebook's shadow accounts are an especially problematic example of how much data we leak. In 2018, Mark Zuckerburg testified to a congressional hearing that Facebook routinely builds accounts on people who do not use the service. They do this with cross-referencing the data they can get elsewhere. When your friends upload their contact list, they get names. They can construct your social circle. They can guess at friends who aren't on Facebook, and then they can start building up information about those invisible people based on what gets tracked across sites that hook into Facebook's digital toolkit.

It's not just Facebook either – your digital fingerprint is smeared all over big tech and there's no way of even knowing what it contains. Thus, the 2012 incident where Target mailed pregnancy vouchers to a teenage girl who hadn't yet worked up the courage to tell her father about her situation (Hill, 2012). If someone is tracking you with cookies, they can use your digital fingerprint to associate your browser with your behaviour.

Isn't it funny how you get an urge to visit a foreign country and the next thing you know all your ads are for travel agents? It's not a co-incidence. Your every click is under scrutiny.

In some respect though, this is a good thing. Few could argue at the value to be found if EXIF data helps police arrest a child pornographer – as happened to Donald Post in 2013 (Ramirez et al., 2015). Or if careless photographs of criminals reveal patterns about their activities. Sicilian police in 2014 for example used phone metadata to map out gang activities that were otherwise opaque (Fielding, 2017). The availability of metadata is not, in itself, a negative state. And neither is the increasing role that technology plays in our lives, although as always there is a dangerous game being played out between our rights, our responsibilities, and our expectations of privacy.

That said, this all does have profound implications for that privacy especially as smart devices become more and more integrated into the home environment. The Internet of Things (IoT) has turned every part of our existence into something more quantifiable – and analysable – than any dystopian prophet could have predicted even as recently as 20 years ago. The extent to which modern society has given up its privacy for its convenience is remarkable, and it's starting to show problems. Not just in the metadata, but in the extent to which we are giving up control of our own environments.

In 2019, Ashley LeMay released a harrowing video of a Ring security camera that had been placed in her eight-year-old daughter's bedroom. The video shows a hacker talking to the little girl. 'Who is that?', the girl asks. The hacker replies, 'I'm your best friend. Santa Claus' (Ramaley & Brooks, 2020). The same year, a Ring camera located in the Brown household in Cape Coral was hacked so as to permit racial slurs to be thrown at the family in their own home.

Despite Amazon assurances to the contrary, the smart device Alexa has been shown to record private conversation and even transmit those conversations to other people (Ford & Palmer, 2019). Samsung televisions are acknowledged to record the conversation in your living room (Gray, 2016), and the small print also says that conversation can be sold on to third parties (Ghiglieri et al., 2017). In many respects this mirrors the common help-line refrain, 'your call may be recorded for training purposes'. And why not? The evidence suggests after all that most consumers are willing to take a privacy hit in exchange for the conveniences of an internet-enabled television (Ghiglieri et al., 2017).

Again it is not as simple as saying 'this is bad' because of the fact these technologies are tremendous enablers of life improvements. Smart devices have created accessibility where none was previously available. They have streamlined life's difficulties. And when used properly, they can combat climate change by ensuring efficiency and optimisation of energy usage within a home. Clearly this comes at a cost – particularly when we consider the increasing burden of technological knowledge each person is expected to master – but for many that cost has been shown to be more than worth it for the benefits.

We also have to ask ourselves the question – what is the alternative? Given the interconnectedness of our lives, what can we even do to ensure that people are aware of the implications of their data hygiene? Imagine a world where everyone had to agree, every time, to share data with the devices they use on a regular basis. What would such a world look like?

You don't really have to think all that far before you come to a common example, cookies in your browser. We hear about these all the time, but it's often unclear what they are. In their simplest terms, they are little packets of data stored on your computer when you visit a website. They may contain things like your user ID, your interaction history with the site, or realistically anything else. The protocols that drive the web are known as 'stateless protocols', which means that every time you visit a website it's as if it was for the first time. The solution to this has been to employ cookies. Our browser is an unfamiliar face at a party, and the cookie is a 'the story so far' document we provide to the host to remind them of all their previous interactions.

Cookies have, for a long time, been at the vanguard of public concerns about privacy on the internet because of the ways in which they have been used. Websites themselves are often made up of resources from multiple different parts of the internet. In order to track user stats, for example, many websites will use an analytical package that gets called in every time a page is loaded. They may draw in images from other locations or build pages in real time from code that is working invisibly in the background. And in some cases, those resources might be things like the Facebook pixel we discussed above. The cookies on your computer, in other words, don't always come from the site you are visiting. So-called 'third-party cookies' have a providence of which you, as a user, may not be aware. Advertisements that follow you around the internet, for example, are almost always a consequence of third-party cookies working in conjunction with the digital fingerprints we leave behind (Eckersley, 2010).

Understanding user concerns about this, the EU included provisions regarding cookies in the GDPR. Specifically, websites must provide users the opportunity to accept or reject cookies when they visit a site. Explicit consent is now required, even if that is difficult to regulate in a way that lends the legislation any teeth (Nguyen et al., 2021).

It seems like an ideal solution, empowering users to decide for themselves what they trust, and do not trust, online. However, in real terms what it has done is create a web environment where the once-derided 'advertising popup' has been resurrected. And in a form that is almost unavoidable unless you delve into the world of plug-ins or the depths of your browser settings. That in turn requires a degree of computer literacy that many users, particularly older users, simply do not possess (Heron et al., 2013) and a willingness on content producers to refrain from the dark patterns that obfuscate understanding (Soe et al., 2020). That the popup is ostensibly for our protection has not reduced the irritation it has created (Giese & Stabauer, 2022; Kulyk et al., 2020).

We find ourselves then in an unfortunate situation. We are swamped with so much information we fall into a kind of choice fatigue (Bauer et al., 2021) that results in us tuning out and simply choosing the path of least resistance with predictable consequences. However, if we are not given the opportunity to consent on a case-by-case basis, we have other issues in terms of being uninformed about where our data is created, used, and sold. And make no mistake – our data is valuable. Facebook alone collected $82 for each of its American users in 2017. This was a sum that it raised largely through selling their data on to advertisers through powerful targeting tools that let campaigns reach individuals that match very precise segmenting needs. Do you need a campaign that reaches women in their 30s with an interest in technology and gaming and with demonstrated interest in left-wing causes? No problem. Here's your bill.

The secondary cost of this is in terms of how much we can trust those that hold our data. Data breaches are common, as we have already discussed in this book. Malicious abuse too is rife – we saw in the 2010s that vast amounts of user data were harvested by firms such as Cambridge Analytica (Isaak & Hanna, 2018). They then built this data into powerful analytical and targeting tools that were sold on to political actors and perhaps even the agents of unfriendly states.

In the scheme of things, perhaps the odd EXIF mishap isn't really where we should be directing our anxieties.

SOCIAL MEDIA POLICIES

There's another topic too that is connected to all of this, and it ties back into what we discussed in a previous chapter with regard to brand management. Universities have a responsibility to police their reputation and often employ specialists and analytical tools in service of that goal. One might even think of this as a social good that is for the benefit of all students – after all, the value of a degree derives at least partially from the prestige of the institute that awards it. Simply having a degree from Cambridge, Oxford, Harvard, or MIT is often more valuable in terms of social capital than any knowledge the studies themselves imparted. As such, if a university allows its name to be degraded it will have a tangible knock-on effect for all staff and alumni.

Organisations all over the world now regularly patrol social media for mentions of specific keywords. After ten frustrated and fruitless emails to a corporation asking them to honour their contractual obligations to a consumer, it's often galling to see a single angry public tweet do better than 'working within the system' ever could. 'I'm sorry to hear that, please send a direct message so that we can resolve this issue' will be the almost instant response that the visibility of social media can bring your way.

That's not to say that the customer support via social media is better – just that it is more public and thus companies are incentivised to make sure they **appear** to be responsive. The request to 'take it to DMs' is not for efficiency of communication, not entirely. It's to move your public complaints once again into a private domain. Corporate social media accounts are railway junctions, shunting oncoming train-wrecks into quiet sidings that have few, if any, witnesses.

You don't even need to tag in the company in a lot of cases. Social media tools allow for companies to track mentions of their brand across the entirety of the open internet even if you made no attempt to directly include them in the tweet or post. Just mentioning them by name can be enough to turn the corporate eye of Sauron your way.

If you could have any superpower, what would it be? We don't know about you, but the one we most **wouldn't** want would be telepathy. We don't want to know what people are thinking about us. We only barely want to be aware of what they are **saying** about us. For the authors, and many others, 'googling' ourselves is a risky proposition. Maybe you'll learn lots of lovely things about yourself. For some though, it's like turning up to a party in your honour held by people that hate you. It's best in many cases to avoid such self-destructive online activities.

Organisations don't have that luxury – they need to massage the public perception of their brands and to make sure they are seen as responsive to public pressure. And on a punitive level they need to ensure that libelous claims or confidentiality breaches are effectively policed. There are considerable consequences for a company that does not protect itself in this respect. And it may be inevitable, in those circumstances, that they attempt to minimise the risk where they have means of control. And one of those means of control is in contract compliance.

An anonymous student in our scandal points out that if she didn't sign the Social Media Usage Policy for the university she wouldn't have been allowed to register. If she couldn't register, she couldn't get a degree. In such circumstances, there's no real option at all. It's coerced agreement where the power differentials are too great to overcome. It's not as simple as 'Well, just go somewhere else'. That's fine if you're choosing between washing machines or mobile phone contracts. You have many alternatives.

The choice to go to a university is bound up in hundreds of considerations and it has life-long implications. People may choose to go to a university because of its location – either close to family obligations or blissfully far away. They may choose it on the basis of its standing in the career path they wish to follow. They may choose a university because it's the one their friends are going to, or because family members have a particular connection. And it may be only one of a few options they have thanks to the seat limit on particular programmes of study. And, inevitably, they're **all** going to have policies like this in place. In such circumstances, it may be an illusion of a choice to say 'Well, you don't have to study **here**'.

We've already discussed somewhat about how social media policies have become a common tool of controlling opinion online. Staff are almost always bound by policies of this nature in one form or another. They place limits – sometimes perfectly fair and appropriate – on what individuals may say on their private social media profiles. For some platforms, such as Facebook, that can be difficult due to the nature of private versus public profiles and posts. For other platforms, such as Twitter, it's trivial. One way to stop people complaining is to put penalties in place for when they do.

It doesn't always work, of course.

The Streisand effect is an internet phenomenon that describes how attempts to divert attention away from a topic will instead intensify the attention that topic receives. The name stems from an incident in which the Oscar winning actress attempted to have some photographs of her coastal residence expunged from a research project aimed at exploring issues of erosion. Instead of protecting her privacy, the incident instead ensured her residence became part of the background culture of internet discourse. Any time someone is caught trying to deflect attention away from something, we all think once more of Barbra Streisand and her luxury mansion.

So when the Union Street Guest House in New York put in place a policy charging guests $500 for leaving a negative review, it doesn't take a genius to guess what happened next. Once *Reddit*, the *New York Post*, and *Time Magazine* brought attention to the policy, it resulted in over 3,000 blisteringly negative reviews. The owners claimed the policy was a joke and eventually removed it from their policy page. The extent to which that joke policy had been enforced, as they claimed it hadn't been, is debated (Cordato, 2014).

Corporations must then walk a tricky path between protecting themselves from negative publicity while also navigating the complexities of internet culture. And like anything, some have seized on the opportunity presented by this as an effective route to publication. Weaponising the effect is now part of marketing handbooks the world over. Do you want everyone on the internet to talk about your product for a while? Then get it censored and the Streisand Effect will do the rest. Just one news story about how PETA's latest shockvertisement has been banned can lead to tens of millions of views on the corresponding YouTube video.

The internet is resistant to censorship. As was once remarked in the early days of the technology, 'The Internet views censorship as damage, and routes around it'. At the click of a mouse, almost anything can be locally preserved. Tweets and social media updates can be removed, but screenshots last forever. On one hand, this is a wonderful fact – the past cannot be rewritten or excised. On the other, the internet is also particularly poor at encoding **context** in its artefacts. Those that wield the past as a weapon have come to cynically rely on the effectiveness of a choice tweet, screen-capped out of context, and revisited at a time of maximum impact.

This decentralisation though is a powerful tool, and one that has been wielded effectively by organisations we have already discussed. As part of its self-preservation strategy, WikiLeaks will release heavily encrypted 'torrents' of leaked data. A torrent is a file that is downloaded from dozens, hundreds, or even thousands of computers at a time. The more people that have a file, the more computers it can be downloaded from, and the faster the download becomes. Everyone shares a bit of the responsibility of distributing it. Since the file is stored effectively everywhere, it's the same as it not being stored anywhere – it becomes impossible to prevent its transmission. Part of the WikiLeaks strategy is that if anything happens to their central infrastructure they only need to transmit a decryption code and the entire world has access to the raw, unedited, and maximally damaging original data. One might think of it as the data security equivalent of Mutually Assured Destruction. It's perhaps the perfect mass-damage interpretation of the Streisand effect. On the Internet, a sense of control – especially of information – is often little more than a delusion.

REFERENCES

Allan, S., & Anden-Papadopoulos, K. (2010). "Come on, let us shoot!": WikiLeaks and the cultures of militarization. *TOPIA: Canadian Journal of Cultural Studies, 23*, 244–253.

Barnes, L. B. (2014). The changing face of espionage: Modern times call for amending the espionage act. *McGeorge Law Review, 46*, 511.

Bauer, J. M., Bergstrøm, R., & Foss-Madsen, R. (2021). Are you sure, you want a cookie?–The effects of choice architecture on users' decisions about sharing private online data. *Computers in Human Behaviour, 120*, 106729.

Benkler, Y. (2011). A free irresponsible press: Wikileaks and the battle over the soul of the networked fourth estate. *Harvard Civil Rights-Civil Liberties Law Review, 46*(2), 311–397.

Berghel, H. (2013). Through the prism darkly. *Computer, 46*(7), 86–90.

Brevini, B. (2017). WikiLeaks: Between disclosure and whistle-blowing in digital times. *Sociology Compass, 11*(3), e12457.

Butler, D. (2016). Tomorrow's world: Technological change is accelerating today at an unprecedented speed and could create a world we can barely begin to imagine. *Nature, 530*(7591), 398–402.

Cordato, A. J. (2014). Can TripAdvisor reviews be trusted. *Travel Law Quarterly, 6*(3), 257–263.

Eckersley, P. (2010). How unique is your web browser? In M. J. Atallah, & N. J. Hopper (Eds.), *Privacy enhancing technologies. PETS 2010. Lecture notes in computer science* (Vol. 6205). Springer. https://doi.org/10.1007/978-3-642-14527-8_1

Eeckhout, L. (2017). Is Moore's law slowing down? What's next? *IEEE Micro, 37*(4), 4–5.

Epstein, R. G. (1994). The case of the killer robot (part 1). *ACM SIGCAS Computers and Society, 24*(3), 20–28.

Fielding, N. G. (2017). The shaping of covert social networks: Isolating the effects of secrecy. *Trends in Organized Crime, 20*(1), 16–30.

Flamm, K. (2018). *Measuring Moore's law: Evidence from price, cost, and quality indexes* (Tech. Rep.). National Bureau of Economic Research.

Ford, M., & Palmer, W. (2019). Alexa, are you listening to me? An analysis of Alexa voice service network traffic. *Personal and Ubiquitous Computing, 23*(1), 67–79.

Ghiglieri, M., Volkamer, M., & Renaud, K. (2017). Exploring consumers' attitudes of smart TV related privacy risks. In T. Tryfonas (Eds.), *Human Aspects of Information Security, Privacy and Trust. HAS 2017. Lecture notes in computer science* (Vol. 10292). Springer. https://doi.org/10.1007/978-3-319-58460-7_45

Giese, J., & Stabauer, M. (2022). Factors that influence cookie acceptance. In *International Conference on Human-Computer Interaction* (pp. 272–285).

Gray, S. (2016). *Always on: Privacy implications of microphone-enabled devices* (pp. 1–10). Future of Privacy Forum.

Heron, M., Hanson, V. L., & Ricketts, I. W. (2013). Accessibility support for older adults with the access framework. *International Journal of Human–Computer Interaction, 29*(11), 702–716.

Heron, M. J. (2016) 'Ethics'. In *Encyclopedia of computer science and technology*. CRC Press. https://www.routledgehandbooks.com/doi/10.1081/E-ECST2-120054032

Hill, K. (2012). *How target figured out a teen girl was pregnant before her father did*. Forbes, Inc.

Hindman, E. B., & Thomas, R. J. (2014). When old and new media collide: The case of WikiLeaks. *New Media & Society, 16*(4), 541–558.

Ioffe, J. (2017). The secret correspondence between Donald Trump Jr. and WikiLeaks. *The Atlantic, 13*.

Isaak, J., & Hanna, M. J. (2018). User data privacy: Facebook, Cambridge Analytica, and privacy protection. *Computer, 51*(8), 56–59.

Karpf, D. (2012). Social science research methods in internet time. *Information, Communication & Society, 15*(5), 639–661.

Kulyk, O., Gerber, N., Hilt, A., & Volkamer, M. (2020). Has the GDPR hype affected users' reaction to cookie disclaimers? *Journal of Cybersecurity, 6*(1), tyaa022.

Kutylowski, M., Lauks-Dutka, A., & Yung, M. (2020). GDPR–Challenges for reconciling legal rules with technical reality. In L. Chen, N. Li, K. Liang, & S. Schneider (Eds.), *European Symposium on Research in Computer Security* (pp. 736–755). Springer.

Landau, S. (2013). Making sense from Snowden: What's significant in the NSA surveillance revelations. *IEEE Security & Privacy, 11*(4), 54–63.

Leadbeater, C. (2020). Living on thin air. In *The information society reader* (pp. 21–30). Routledge.

Lynch, L. (2013). The leak heard round the world? Cablegate in the evolving global mediascape. In *Beyond WikiLeaks* (pp. 56–77). Springer.

Macewan, N., et al. (2008). The Computer Misuse Act 1990: Lessons from its past and predictions for its future. *Criminal Law Review, 12*, 955–967.

Macken, C. (2005). Preventive detention and the right of personal liberty and security under the international covenant on civil and political rights, 1966. *The Adelaide Law Review 26*(1), 1–28.

Madhisetty, S., Williams, M.-A., Massy-Greene, J., Franco, L., & El Khoury, M. (2019). How to manage privacy in photos after publication. In *Proceedings of the 21st International Conference on Enterprise Information Systems* (Vol. 2, pp. 162–168,). *ICEIS*. ISBN 978-989-758-372-8; ISSN 2184–4992, SciTePress. https://doi.org/10.5220/0007614001620168

Mahoney, P. (2004). Right to a fair trial in criminal matters under Article 6 ECHR. *Judicial Studies Institute Journal, 4*(2), 107–129.

Masco, J. (2017). 'Boundless informant': Insecurity in the age of ubiquitous surveillance. *Anthropological Theory, 17*(3), 382–403.

Montasari, R., Carroll, F., Mitchell, I., Hara, S., & Bolton-King, R. (Eds.). (2022). *Privacy, security and forensics in the Internet of Things (IoT)* (No. 1). Springer.

Montasari, R., Peltola, P., & Carpenter, V. (2016). Gauging the effectiveness of computer misuse act in dealing with cybercrimes. In *2016 International Conference on Cyber Security and Protection of Digital Services (Cyber Security)*, IEEE, London (pp. 1–5). https://doi.org/10.1109/CyberSecPODS.2016.7502346

Nguyen, T. T., Backes, M., Marnau, N., & Stock, B. (2021). Share first, ask later (or never?) Studying violations of {GDPR's} explicit consent in android apps. In *30th USENIX Security Symposium (USENIX Security 21)* (pp. 3667–3684). USENIX Association, California.

O'Loughlin, J., Witmer, F. D., Linke, A. M., & Thorwardson, N. (2010). Peering into the fog of war: The geography of the WikiLeaks Afghanistan war logs, 2004–2009. *Eurasian Geography and Economics, 51*(4), 472–495.

Ramaley, K. J., & Brooks, B. W. (2020). Ring security: Peace of mind vs. chilling nightmare. *Journal of Critical Incidents, 13*, 89–92.

Ramirez, R., King, K., & Ding, L. (2015). Location-location-location-data technologies and the fourth amendment. *Criminal Justice, 30*, 19.

Sarigol, E., Garcia, D., & Schweitzer, F. (2014). Online privacy as a collective phenomenon. In *Proceedings of the Second ACM Conference on Online Social Networks* (pp. 95–106). Association for Computing Machinery, New York, NY. https://doi.org/10.1145/2660460.2660470

Scheuerman, W. E. (2014). Whistleblowing as civil disobedience: The case of Edward Snowden. *Philosophy & Social Criticism, 40*(7), 609–628.

Soe, T. H., Nordberg, O. E., Guribye, F., & Slavkovik, M. (2020). Circumvention by design-dark patterns in cookie consent for online news outlets. In *Proceedings of the 11th Nordic Conference on Human-Computer Interaction: Shaping Experiences, Shaping Society* (pp. 1–12).

Toffler, A. (1970). *Future shock*. Pan.

Tufekci, Z. (2018). Facebook's surveillance machine. *The New York Times, 19.*

Walsh, P. F., & Miller, S. (2016). Rethinking 'five eyes' security intelligence collection policies and practice post Snowden. *Intelligence and National Security, 31*(3), 345–368.

Worthy, J., & Fanning, M. (2007). Denial-of-service: Plugging the legal loopholes? *Computer Law & Security Review, 23*(2), 194–198.

Zeman, J. (2015). A slender reed upon which to rely: Amending the espionage act to protect whistleblowers. *Wayne Law Review, 61*, 149.

CHAPTER 14

Clean-Out at Scandal-Linked Journals

NEWSPAPER ARTICLE

An exclusive report by Jack McKracken for the Dunglen Chronicle

Several journals which had published work from Professor Blackbriar of the University of Dunglen have been quietly cleaning house in recent weeks. Many members of their editorial boards have been suspended and several peer reviewers struck off internal registers and blacklisted. This according to Dr. Paula McCrane. Dr. McCrane was previously a deputy editor for the Journal of Deep Sea Oil Exploration (JoDSO) and is currently editor-in-chief of a new open access academic outlet called the International Journal of Extreme Extraction (IJEE).

'I received an email from the owners of JoDSO a few weeks ago that said, in essence, "this happened under your watch, and so we won't be requiring your services any more". I'm not the only one who got an email like that'.

'The process of academic publishing is misunderstood by many', she added. 'My position as deputy editor was unremunerated – I did it just because it's part of what it means to be an academic. I handled correspondence with contributing authors, chased up calls for papers, and allocated peer reviewers to submitted papers. I also did the "desk reject" if necessary – that's when we say we won't publish a paper and don't even send it out to reviewers. My role in the publication of the Blackbriar papers was really just logistical, but that doesn't matter it seems'.

The Journal of Deep Sea Oil Exploration has come under intense criticism by academics since the

DOI: 10.1201/9781003426172-14

scandal broke. Many have bemoaned the decision by the journal to accept papers that did not clearly outline the methodology used to select data for analysis. Others have pointed out that the institution subscription fee of £6,000 a year means that the research is often hidden behind paywalls and that this discourages other academics and experts from applying the requisite amount of scrutiny.

'Peer review is often lauded by outsiders as the gold standard of academic research', says Professor Callum Sutherland at the University of Alba, 'But people often don't know what it involves. It doesn't mean that the paper's experiments are reproduced. It just means that people who know about the field have read the paper, feel that the conclusions are justified by the written analysis, and that there is suitable originality to warrant publication. Peer review relies very heavily on trust – if there are actual falsehoods in a paper, they'll only be found if they fail a "smell test". That is, if they contradict other papers, or report on results that are wildly out of expectation. Bear in mind in the Blackbriar case that the professor had been reporting incremental gains in his research for decades. There was no sudden leap from modest to mind-blowing in terms of the scale of his results'.

The increased scrutiny applied to the journals as a result of the scandal has shone a light on several other malpractices. These are likely to result in retractions across the whole academic catalogue of the affected publishers. AI analysis of text revealed several incidents of plagiarism, along with many examples of misquoted references. Also problematically, there is evidence of an exponential increase, year on year, of papers that include substantive portions of computer-generated text. Close examination of those papers revealed a number of fake references employed in the support of spurious argumentation. The value of a journal lies in the robustness of its contents, and it seems that this robustness is now seriously in question.

Universities across the sector are having to cut back on journal provision for their academics as the price for subscriptions continues to increase. 'We spend around £4m every year on journal subscriptions', says University of Dunglen librarian Bill Rising. 'And that cost keeps going up. Some of the most significant and important journals cost £10–20k each for a year. If we buy individual articles rather than subscriptions, it can easily cost between £30 and £100 for each. As such, we can't just subscribe to everything – we need to maximise the effect of our spending, and some journals just aren't as useful to us as others. Other universities are faced with the same challenges, and that means that academics often don't have access to scholarly literature. That limits their ability to fully participate in the academic process'.

Dr. McCrane explains her solution to this problem. 'The issue is that the whole process, start to finish, for many journals is entirely voluntary.

People submit their papers for free. Editors deal with the administration for free. Peer reviewers peer-review, again for free. There are few, if any, real staffing costs. So that's why I started an Open Access journal – one where we do not charge people who wish to read the work, all the costs are borne up front by either the journal or the author. It does mean that sometimes we need people to pay to publish their papers, but that money comes out of research grants and the overall effect is a huge positive for academia'.

Peer review is normally a blind process, where those submitting papers receive only the feedback, and not the names, from those who have critiqued their work. The editor handling the publication knows, and records are kept to ensure that there is a fair distribution of responsibility and that no conflict of interest is permitted.

Responding to comments by Dr. McCrane relating to her role in the paper publication, the editor of the Journal of Deep Sea Oil Exploration Dr. David Sumner said, 'I'm very sad that we have had to remove Paula from the editorial board. It was though her responsibility to ensure the peer review was conducted ethically. Unfortunately, among the five peer reviewers selected for the last Blackbriar paper was Mrs. Shona McAlpine – his PhD student. She sent back a feedback sheet that indicated there were no problems with the paper and that it should have been accepted immediately for publication without corrections. Similarly, the last paper co-authored between Blackbriar and Mrs. McAlpine was peer-reviewed by James Duncan, another PhD student of the professor. This is a clear conflict of interest and strikes at the heart of the integrity of our peer review process'.

Dr. McCrane responds by saying, 'In very specialised research fields, we're often required to request review from people who have connections with other people. Everyone knows everyone in some fields, and you can't find many people who are both experts and unconnected to the author. Similarly, we sometimes send out peer review requests to previous authors who haven't yet attained their doctorates – we're interested in expertise, not credentials. However, we did know that both James Duncan and Sharon McAlpine were former collaborators with Professor Blackbriar, and their feedback was weighted accordingly. They were each only one of five, after all'.

We spoke to one blacklisted peer reviewer, who wished to remain anonymous. He said, 'I've published a couple of papers but I'm still in the process of building my reputation in the field. I had high hopes of maybe getting to work with Professor Blackbriar at some point. While peer review feedback is anonymous people have styles of writing and patterns of expertise and expectations that are like a signature. I didn't want to get on his bad side, and I knew that he'd published dozens of times on this topic before. Nothing about the paper looked suspect, so I recommended its publication'.

Callum Sunderland adds, 'Universities are unusual in that they remain, even now, mostly a gift culture. You're not valued by what you have, but rather what you give away – in this case, the papers that you publish, and the credibility you afford the papers you review. Academics don't profit financially from the massive costs journals impose on libraries, all they get from a paper publication is recognition and respect. That respect generates collaboration opportunities and justifies research funding coming their way. The financial incentives of the journals, the self-correcting mechanics of science, and the reputational incentives of the academics involved often do not seamlessly align'.

Emeritus Professor Joanne Clement, who took early retirement from the University of Dunglen, makes a different argument. 'Academia isn't a gift economy. It's a rejection economy. Ninety percent of what you are told is "no". No, we won't publish that paper. No, we won't fund this grant. No, you can't have that job. If a gift economy is marked by the value of what you give away, academia is marked by the value of what people choose to accept from what you offer. Journals, conferences and funding bodies set great store by their rejection rate – getting a paper accepted where only ten percent make the cut is much more prestigious than one where fifty percent are accepted, right? That paper, by definition, must be of higher quality, yes? Of course, it's not – the role of journals in ensuring quality is overblown. They have a vested interest in ensuring lots of submissions, which leads to more punitive acceptance rates. Peer review is ninety percent gatekeeping and ten percent standards-keeping. Many disciplines have chosen to completely bypass the whole system and work primarily through preprints now. That way you can focus on the scholarship, and not the brinksmanship. It's the perceived value of rejection that creates the perceived value of published research'.

Two senior engineers at ScotOil have hit out at the cost of access to research. 'Our company paid for Blackbriar's research, and while we get access to the raw results of his studies we don't get to see the actual published papers. We just see pre-prints and early drafts, unless we pay money for the research outputs we funded. Some of us may have had queries about the results if we'd been able to see the same papers that everyone else was reading. We know the North Sea better than anyone, but we're priced out of participation in the review of scientific studies based in the area. We outsource a lot of our research and development to academia. We can buy the odd paper here and there, but we can't afford to spend hundreds of thousands a year on getting access to all the journals in this area on the off-chance there's something published that's relevant to us'.

Professor Blackbriar was unavailable for comment.

THE ECONOMICS OF THE ACADEMIC GIFT CULTURE

We have focused heavily within this scandal on the reputational damage to individuals. John Blackbriar, Sharon Macalpine, James Duncan, and many more will carry this incident with them long after its relevance fades. But a scandal like this has impact beyond that of individual people, and it strikes at the heart of the academic system itself. It's easy for those outside of academia to be confused about how it all works, because it can be mystifying to outside observers. As we've already pointed out, when one of us remarks to a parent 'I had a paper published this week', that parent will almost always ask, 'And how much do you get paid for that?'.

It would be even more baffling if we explained that often **we** need to pay for a paper to be published – at least, if we want it to be open access. And if you break the academic publication system down, it does indeed seem like a system designed by aliens.

1. Taxpayer money is used to fund academic time – either as part of a general 'research' allocation or through specific research grants.
2. That time is used to conduct research, run studies, and analyse the results.
3. The most interesting (ideally) parts of that work are then encoded into a paper, which is written, edited, and often typeset by the academics that conducted the research work.
4. The paper is submitted to a conference or a journal, where it is peer-reviewed by other academics on an unfunded basis.
5. The paper is (hopefully) accepted for publication, at which point the taxpayers that paid for it are charged to access it when it is placed on a secured site. This is known as a paywall. Universities will usually pay to subscribe to a large number of journals, allowing their academics to read the paper for 'free'. The general public? No, pay up.
6. By paying an average Article Processing Charge (APC) of around $2000, the article can sometimes be made free for everyone to view.

In some cases subscription charges to libraries are going up at exactly the same time that conventional publishers are shifting to open access. In effect, we're seeing not only public money funding private profits but a double-dipping in terms of APCs and institutional subscriptions.

The academics, peer reviewers, and (usually) editors are not paid anything for participating in this process. Whenever someone pays for access to one of our articles, nobody who did the work gets a penny and certainly none of it goes back to the taxpayer that funded it. Indeed, the taxpayer is double charged in most cases – once for the work to be done and then again for the privilege to read the results. That's even true of the academics that wrote the paper – if their university doesn't subscribe to the journal, even they may have to pay to access their own past work.

Where **does** all this money go? Well, it goes into the coffers of the multinational publishing houses that own the journals. And business is booming. Companies like Elsevier, Black and Wiley, Taylor and Francis, Springer Nature, and SAGE control over half of the academic publishing market (Lariviere et al., 2015). That was worth almost $20 billion in 2022. A traditional newspaper – paying a payroll of journalists, editors, artists, and other expenses – may have a profit margin of around 10%. Journal publishers, presiding over a transfer of wealth from the public to private owners, may have a profit margin closer to 40% once all the accounting tricks are factored in (Van Noorden et al., 2013). This waterfall of money is not surprising given how much of their content is simply given to them by academics. As one conservative commentator put it, 'Academic? You're just a cash-hamster spinning a publisher's profit wheel'.[1] It's hard to argue, especially when estimations suggest that peer-reviewing by itself represents an annual 'billion-dollar donation' (Aczel et al., 2021) within the US alone. When you add in the Chinese and UK markets, the total comes to an estimated 2.5bn dollars per year.

As to why this is the case though, we need to look at academic publishing in a different light – that of the 'gift culture'.

In traditional economic structures, worth is partially determined by how much you have, or perhaps the worth of what you create. Your outputs have a financial benefit associated with them, and you are as valuable as the sum of your assets plus your productivity. Conspicuous consumption is a consequence of this model, where the signal of 'I have more wealth than you' is a proxy for 'And thus I am a more valuable person than you'.

Gift cultures work differently. Your value is determined at least in part by the value of what you **give away**. The amount which you have freely contributed to society as a whole serves as the basis of your reputation. As such, to freely provide (or provide at your own cost) a 'contribution to knowledge' is your gift to the world. You do not profit from it, but you pass it on in the intention others will.

Really that's the currency of academia, and the real reason why the consequences are so high for those implicated in the scandal. Academia is a reputation-based economy. Every time your paper is cited, that's a boost to your reputation. As you cite other papers, you give them a little bit of your own reputation by establishing them as important in the construction of your own work. Reputation leads to opportunities. These include promotions; employment in more prestigious institutions; and more credibility when bidding for substantial amounts of research funding. That's the funding which is used to do more research and thus build more reputation.

That all sounds nice and fluffy and represents an idealised view of academia – that of socially minded scholars seeking only to improve the world in which they live. However, realistically, we see here the issue of coercion. The simple fact is, you work within this system because nobody has yet proposed a realistic way of dismantling it. Contentious objection is only possible for those that have already benefited from the system – tenured professors, permanent academic staff, and those that have made management their core function. Everyone else has to simply work within the lines laid out if they want to succeed.

This coercion starts early, right at the doctoral studies stage. If you want to smoothly progress through a PhD and its associated defence, then you better have a few good papers to back up your case. Published papers in well-regarded journals and conferences show that you have uncontroversially met the 'unique contribution to human knowledge' expectation of a PhD degree and thus you are in little danger of failing. And also, for your own career progression, you're going to need them for the next step – getting a job in academia.

Nowadays the path to a permanent academic position – certainly in the UK and the US – tends to go through multiple postdoctoral positions. In these positions you are employed on short-term contracts within fixed-duration research projects. Each of those projects will (hopefully) generate papers, and those papers will be needed to parlay your way up the chain into a lectureship position. The number of papers is important, but so too is the number of **citations**, as is the perceived importance of the venues in which you publish. Your gifts alone won't carry you forward, those gifts need to have been appreciated by those who live within the culture you wish to join.

Your first lectureship may be term limited, and achieving permanence is another barrier which is – you guessed it – determined by your papers and citations. A permanent academic position is often described as one with 'tenure'. Once upon a time, a good research record would have been enough, but what you also need to do now is prove a track-record of 'grantsmanship'. As in – bidding for research money **and** having it awarded. The latter is outside your control, of course… as we have seen, only a minority of bids are funded and the quality of a proposal is only somewhat causatively linked to its success (Bol et al., 2018; Tamblyn et al., 2018). The best way to get a good amount of research funding is to bid for **a lot** of research funding, and to bid for it often. And all the way along the line you are producing profit for the academic publishers that have a vested interest in you continuing to operate their profit wheel.

How do you break out of this system? You don't, until you've reached a point of sufficient safety and seniority where you can say 'No more'. By then, you've likely already internalised the message of 'publish or perish'. Those most in a position to break the chains are the ones most likely to be willingly struggling beneath them.

We see here that Dr. Paula McCrane and a roster of peer reviewers are being suspended and struck off journal teams for malpractice. What we **should** see is massively profitable corporations punishing those that have freely provided the services that fuel their massive profits. The journal owners, in scandals such as these, escape unscathed. Not only do they draw in all the profits, they manage to side-step all the criticisms… such as with regard to the role they play in enabling a system where academics feel compelled to engage in fraudulent practices. And when all you get from working within this system is reputation, it stings all the worse to have that reputation ruined.

Perhaps to cast it in somewhat Marxist terms – academics often feel alienated from the products of their labour. Few academics support this system, at least in theory. Few academics believe paywalls are ethically or morally appropriate. There's just no obvious way to jump off of the profit wheel while so many other hamsters are scurrying away.

THE REJECTION ECONOMY

Clement, in our article, makes a slightly different claim – that academia is not a gift economy. Rather, it is a rejection economy. Those on the inside will know just how true that is. We've spoken about the low funding rates for grants. Ninety percent of the time you'll be told no, even after months of intensive labour building a consortium and developing a programme of research. We've spoken about rejection rates for papers. Again, for a 'high-impact' journal or conference you might be looking at a lot more rejection than acceptance. CHI 2023, one of the most prominent conferences in human-computer interaction, had a 27.6% acceptance rate. In some recent years that has been as low as 20%. Nature, one of the most prominent journals for the natural sciences, has an 8% acceptance rate. Many other journals will hover around the 30% mark. Again – most of what people hear upon submitting a journal or conference paper is 'no'. We hear 'no' when applying for jobs, where supply as we have discussed massively exceeds demands. Most of our 'scholarly' activities are done to a backdrop of rejection.

The perceived value that we give to the world then is in many ways correlated with just how hard other things were rejected. Publishing at CHI is more prestigious, in the field, than publishing in a local conference. How could it not be, when your paper survived a gauntlet that crushed so many others? The truth, as it often is, is more complex than that. Many conferences/journals are defined by their 'impact factor', which is to say the average number of papers that cite its papers in a year. If a conference has an impact factor of ten, it means on average each paper at the conference gets ten citations. That's a lot. You'd think impact factor correlates tightly with rejection rates, but it's only somewhat true and the effect is difficult to observe across disciplines (Bjork & Catani, 2016). In short, while there is certainly academic prestige to be gained by publishing in outlets with a high rejection rate, it doesn't seem to effectively translate into actual **scholarly impact**. It's possible to publish in only the best journals and conferences and still receive few, if any, citations. The opposite is true too – a successful academic career can come from publishing in the outlets that other people actually read, as opposed to the places where people are mainly interested in seeing their own writing.

Nonetheless, academics every day fight the stats to get their work out there because it's how the system is described to those trapped within it. And realistically, almost everything in life that we associate with merit is the product of a rejection economy. This isn't unique to academia.

We value bands that make it through the AR process of a label. We provide greater cultural cachet to books that are published as opposed to self-hosted fan fiction. We value authentic awards over vanity awards. Rejection serves a powerful role in curating attention – in its best forms what it does is highlight the best at the expense of the worst. However, it is foolish to assume that poor quality is the main causative factor associated with rejection. The Beatles were rejected by Decca because 'guitar bands were on the way out'. *A Confederacy of Dunces* by

John Kennedy Toole was considered unpublishable by the author. It was eventually published to great acclaim 11 years after he died. *Harry Potter and the Chamber of Secrets* was rejected by 12 publishers until Bloomsbury realised its promise. Perseverance in the face of rejection is one of the critical skills that must be developed by anyone working in a rejection economy.

That said, perhaps the situation is sadder for academics given the discussion above regarding the economics of academia. Being rejected for something you're trying to give away for free, to the profit of a publishing company, feels somewhat sharper than someone hoping their album is going to be a hit big enough to leave them drowning in fans and money and fame. Joanne Clement may sound bitter. In that she's perhaps just channelling the frustrations of an entire sector.

With all this in mind, perhaps we can see slightly better why sometimes things fall through the cracks. For many years, there have been experiments in submitting ridiculous papers to conferences and journals. The intent is to see if reviewers can genuinely tell the difference between literal nonsense and the academic doublespeak required to convince a reader of 'scholarly merit'. These stings have varied motivations (Teixeira da Silva, 2021) and are rarely received positively by those who are most committed to a perception of academic rigour in publishing (Al-Khatib & Teixeira da Silva, 2016; Grech, 2019). That's fair. Those with an identity most invested in the rejection economy are the ones least likely to throw suspicion upon its foundations. Careers can be built upon successful mastery of the game of academic publishing, and few are willing to argue that the game might be rigged, with systemically broken rules.

That said, critics of these stings are quite correct in that much academic publishing works on a trust model… but that seems at odds with the perception often associated with the corrective values of peer review. Peer review, we are told, highlights methodological problems, faulty logic, and spurious conclusions. Sometimes though it seems to do nothing of the sort. Hoax articles – whether written by humans or by AI – are rarely submitted in a systematic way that would support evidence of widespread lax standards in academia (Lagerspetz, 2021). They tend instead to be laser-targeted on those outlets most likely to support an ideological argument. That is not to say they don't reveal uncomfortable truths of what happens at the periphery of our disciplines, or disciplines critical to the construction of public narratives around social issues.[2] It's just that such truths are difficult to convincingly extrapolate to the entire sector.

You don't really need to go to this length to get a paper published these days though. Some disciplines have given themselves over to pre-prints as the primary mechanism of distributing early and potentially groundbreaking work. These are academic papers, put online prior to formal peer review, which then later might be submitted to a formal journal. They serve as a way of putting a stamp of ownership on an idea and allowing the community as a whole to collaborate on late-breaking work – this obviates the need for the many-month publication cycle of even snappy journals. If the promotion structures of these disciplines align with the perceived value of pre-prints, it can be a way to focus more on the science than the process. For those only interested in the illusion of participating with the rejection economy, there are plenty of predatory journals out there that will publish anything you want provided you don't mind contributing a 'small processing fee'.

Coupled to all of this, we also have to harken back to an earlier article in which Sharon McAlpine was noted to have produced some of the text of a paper with the help of generative AI. Journals are increasingly taking a hard line in this, with many stating openly that co-authorship of a paper with ChatGPT and other tools is considered a violation of their academic standards. It is likely that, given the scrutiny this scandal will have brought to all the papers in Blackbriar's bibliography, that at least one of the publications with which she was involved may require a messy and embarrassing retraction.

NOTES

1 https://conservationbytes.com/2019/09/09/academic-youre-just-a-cash-hamster-spinning-a-publishers-profit-wheel/
2 https://en.wikipedia.org/wiki/Grievance˙studies˙affair

REFERENCES

Aczel, B., Szaszi, B., & Holcombe, A. O. (2021). A billion-dollar donation: Estimating the cost of researchers' time spent on peer review. *Research Integrity and Peer Review*, *6*(1), 1–8.

Al-Khatib, A., & Teixeira da Silva, J. A. (2016). Stings, hoaxes and irony breach the trust inherent in scientific publishing. *Publishing Research Quarterly*, *32*, 208–219.

Bjork, B.-C., & Catani, P. (2016). Peer review in megajournals compared with traditional scholarly journals: Does it make a difference? *Learned Publishing*, *29*(1), 9–12.

Bol, T., de Vaan, M., & van de Rijt, A. (2018). The Matthew effect in science funding. *Proceedings of the National Academy of Sciences, 115*(19), 4887–4890.

Grech, V. (2019). Write a Scientific Paper (WASP): Academic hoax and fraud. *Early Human Development, 129*, 87–89.

Lagerspetz, M. (2021). "The grievance studies affair" project: Reconstructing and assessing the experimental design. *Science, Technology, & Human Values, 46*(2), 402–424.

Lariviere, V., Haustein, S., & Mongeon, P. (2015). The oligopoly of academic publishers in the digital era. *PLoS One, 10*(6), e0127502.

Tamblyn, R., Girard, N., Qian, C. J., & Hanley, J. (2018). Assessment of potential bias in research grant peer review in Canada. *Canadian Medical Association Journal, 190*(16), E489–E499.

Teixeira da Silva, J. A. (2021). Assessing the ethics of stings, including from the prism of guidelines by ethics-promoting organizations (COPE, ICMJE, CSE). *Publishing Research Quarterly, 37*(1), 90–98.

Van Noorden, R., et al. (2013). The true cost of science publishing. *Nature, 495*(7442), 426–429.

CHAPTER 15

Student Journalism Outs Senior Academics

NEWSPAPER ARTICLE

An exclusive report by Jack McKracken for the Dunglen Chronicle

In an incredible scoop regarding the University of Dunglen, citizen journalism has revealed a stunning link between senior members of the university and various funding partners working on the North Sea Algorithmic Exploration (NSAE) project. A PeerNet page set up to support Professor Blackbriar, called 'Professor Not Patsy', has attracted considerable online attention since the scandal broke. It has been the source of gossip, snippets of information, and conspiracy theorising for weeks. Last night however a regular of the page revealed that Sir Gideon Lazenby, university treasurer, was on the board of directors for ScotOil. Professor McManus, chair of the research ethics board, was a key member of the European Funding Council committee who approved funding for Blackbriar's scandal-linked project.

'It's incredible!', said the PeerNet poster revealing this information. 'They lobby for millions of funding they control to go to their institutions, and then they use that very same funding to bully the professor into publishing fake data! If anyone is to blame, it's these bastards! They knew the whole thing was dodgy right from the start!'

The allegations have added new fire under the feet of the university senior management team, with a broad base of calls coming from across the sector for them to resign. 'It's a clear conflict of interest', said one anonymous staff member. 'It's borderline Machiavellian. They

DOI: 10.1201/9781003426172-15

were the ones that made the stick that John ended up beating himself with, and they're also the ones that talked him into picking it up in the first place!'

However, amongst all the legitimate commentary on both sides of the debate we have seen an increase in false claims, deep fakes, distorted memes, and straight-out inventions of fake news. Increasingly, it seems like much of the commentary across Blether is automated in some fashion, with vested interests creating bots that can respond to any expressed interest with content – often manipulated – that skews towards their own talking points.

Correspondence received by this newspaper under a Freedom of Information Act query show that Professor Blackbriar was initially opposed to seeking out funding from the NSAE on account of the fact that his algorithm did not work reliably in all circumstances.

A minute obtained through the act reveals the following entry:

`Professor Blackbriar believes that the early promise shown by his research in finding exploitable natural oil reserves represents a strong research direction, but still requires further calibration to ensure the results are actionable in edge-case scenarios. He believes a further five years of study are required before his toolkit can be used for real-world projects within the NSAE.`

The NSAE was both a research project and commercial endeavour. The charter of the project outlines its goals as follows:

`To apply cutting-edge academic and commercial research to extending the productive life of deep sea drilling rigs, and to use the data generated in the application of this research to further refine the tools used for other drilling environments.`

'The expectation of the NSAE was that all the tools being used in its day-to-day operation were largely "working"', said Vanessa Haynes, spokesperson for the NSAE project, today. 'The research data we were gathering was not to make the tools work, but to validate them and to make them work in other places – not just the bottom of the sea. It was also designed for a dozen or so processes and calculations to work together, each feeding into the other. One badly implemented link in that chain would throw all the following ones out of alignment. Blackbriar's algorithm was in the middle of that chain'.

Minutes from the University show an increasing amount of pressure being applied to Blackbriar as the deadline for an application to participate in the NSAE loomed:

`Professor Sir David Tumblewood asked again whether Professor Blackbriar's algorithm had correctly` identified extractable resources in pilot testing. Blackbriar confirmed that it had, but that these results were not generalizable. Sir David once again highlighted the financial consequences for the university and the looming deadline. Sir David asked if Professor Blackbriar had alternate funding for his department in mind. Professor Blackbriar confessed he did not.

Tumblewood then pointed out that part of the NSAE's remit was to investigate the edge-cases that Blackbriar was most concerned with. Blackbriar agreed to integrate his algorithm into the wider ecosystem of NSAE.

When the application to participate in the project was received, it was put out for review to two key committees. The first of these was the European Funding Council who assessed the likely cost versus eventual benefit, and the likely ethical and environmental considerations.

'Sir Gideon Lazenby was a key member of the funding committee who scrutinised the proposal for the economic and ethical implications', said Chuma Hassan today. 'While his institute did stand to profit from approval, his impeccable reputation for scrutiny for bid applications meant that we did not feel the need for him to recuse himself from the final judgement. He has been on this committee for many years, and never once have we had any doubts about his probity. In light of recent revelations regarding his internal role within the university, we perhaps should have insisted on a replacement for his vote'.

The second committee was the Collaborative Research Committee (CRC) of the NSAE. This was a board of ten senior engineers from the main industry funding partners. The results of this committee were then passed to each of the board of directors for approval. Sir Gideon Lazenby was a member of one of these boards. His board voted to approve the application, with a majority of seven to six.

'It's important to note here that there was some criticism of our involvement in the project that came from the CRC', said Derek Simmons, Vice President of Industry Collaboration for ScotOil. 'It was far from a clean run for the project – it was Lazenby in the end that took it from being a dead tie. A tie which would have meant at least a couple of years delay while the kinks in the project were ironed out. As it is, it narrowly received ascent, and it's looking increasingly likely that ScotOil will have lost fifty or sixty million pounds as a result of that vote'.

The citizen journalism that revealed this story is most remarkable because of the extent to which both the EFC and the CRC have gone to hide the details of their deliberations. The ScotOil committee minutes are located deep within their internal intranets and thus are not available for scrutiny from members of the public.

'These are commercially sensitive documents', says Simmons. 'We can't have them out there for our competitors to pore over. I'm not sure how they got out at all, but we've instituted a full leak inquiry. Initial investigations suggest that someone had leaked the information via the "dark internet", so we can't easily track where it came from. It looks like it was sent over TOR, or one of the systems used by drug-dealers and paedophiles'.

The EFC is a European Union public committee, but shortly after the scandal broke much of the information relating to the project was made inaccessible on their website. 'You'd just get a 404 error when you followed a link', said one journalist who had been investigating. 'The link implied the file was there once, and the 404 error that it had been removed in the interim. Once it's removed it's gone. We had been looking into details of who had been behind the EFC approving the project, but they obviously didn't want us to know. Most of them may have been civil servants, politicians and bureaucrats – they know how to hide their mistakes'.

Not so, according to the participants of the Professor Not Patsy PeerNet group. 'We went to the Internet Archive', said one poster. 'That has copies of pages taken at all kinds of times in the past. The membership of the funding council committee that approved the bid was saved there, and we downloaded it locally. You can't scrub your mistakes off the internet, and you just look bad for trying'.

Professor Blackbriar was unavailable for comment.

CONFLICTS OF INTEREST

The article Student Journalism Outs Senior Academics makes an extraordinary set of claims that carry with them almost unanswerable challenges to the integrity of the university management structure. Specifically:

1. Sir Gideon Lazenby, university treasurer, is a member of the ScotOil board of directors and thus had a financial interest in the results of Blackbriar's research. He also had the ability to influence decisions taken regarding funding.
2. Professor McManus, chair of the research ethics board, was part of the committee of peer reviewers who approved funding for the project in the first place.

The implications here are significant. The NSAE project at the heart of the scandal had the potential to impact everything from annual earnings to stock prices for the ScotOil corporation. Public perception that it was successful was important in managing the latter of these, whereas the actual success of the algorithm is more important for the former. The thing is, while it is obvious that ScotOil benefits in the long term from wise investment and governance, it is not always true that such investment is the optimal path in the short term. We must consider here for example the business model that underlines several major industries where the profit comes from the philosophy 'grow big and then work out how to make money'. Stock price in the short term is often seen as a larger generator of individual wealth than slow, steady but reliable profits (Dallas, 2011; Terry et al., 2015) As such, we see Lazenby here not just as a voice of counsel in the ear of Blackbriar, but someone who may very well now be guilty of insider trading (Jaffe, 1974).

It is not necessarily illegal, or even unethical, for Lazenby to be on the board of a company working with the university – on a broader, societal level it may even be beneficial in some cases (Cho et al., 2017). Directors are often closely associated with external partners, given they represent major stakeholders in ongoing activities. However, here we see a clear conflict of interest (Felo, 2001). Dealing with these is at the core of ethical decision-making.

There is a conflict of interest when the incentives for our desired outcome are misaligned with the incentives of a desired outcome of another involved party and they happen all the time. Conflicts of interest are unavoidable in certain circumstances, such as the one Lazenby finds himself in which he has a financial stake (through ScotOil) in the success of his colleague's work (Blackbriar). Lazenby also though has a stake in authenticity in research (part of his general code of ethical responsibilities as a university academic), and in the financial sustainability of the university (through his role as university treasurer).

The common solution here is for stakeholders to 'disclose their interests', as this is the first step in resolving a conflict of interests – transparency as to where those interests lie. The next step is to remove oneself from all conflicts where that is an issue… typically to 'recuse' themselves from any decision-making and refrain from commenting or providing advice. It is a transparent 'interest' for example for someone to want a friend to get a job in the same company. It is a **conflict** of interest when that person also has decision-making responsibility in that employment, at which point a recusal is appropriate (Mecca et al., 2015).

Professor McManus in turn has a slightly different conflict of interest here, in that he perhaps does not stand to benefit personally (as best we can tell) but is in the position of ruling preferentially on decisions that will greatly advantage his workplace and burnish his own reputation as a result. He also has an interest in ensuring public perception regarding his previous decisions (heading the research ethics board) does not sour.

Dealing with these conflicts is not necessarily difficult – we have ethical systems in place to permit this kind of thing to be done cleanly. It seems likely though that Lazenby has not transparently declared his interests (otherwise it wouldn't be a surprise to anyone) and did not recuse himself from any decisions involving his university when on the ScotOil board. McManus in turn should likely have recused himself from consideration of any potential EFC grant application that came from his university (Lemmens & Freedman, 2000).

As usual, it's not quite that simple. Chuma Hasan points out that Lazenby has an impeccable reputation for scrutiny. Recusal though isn't a tool for removing people because they're biased – people can mentally control to an extent for biases they know exist (Moore et al., 2010; Wilson & Brekke, 1994) although with notoriously unreliable results. Recusal is for making sure that the biases we **don't know** we have don't sneak in. It is likely that Lazenby believed his vote was purely objective and evidence-based. It's entirely possible that it was – those with a reputation for fair-mindedness may value it to the point of being **more** critical of things in which they have a vested interest. It may be that the absence of Lazenby would have left the committee with a knowledge gap that would have resulted in a worse outcome. Recusal may in the end be a luxury in some circumstances, since it requires someone else step up in the place of the person recusing themselves. Maybe there was no one available. All of these things are perfectly possible, and perhaps even probable. However, when things go bad, they are justifications that ring singularly hollow.

When such conflicts cannot be dealt with, because they involve no clear path through which all interests can be satisfied, then such conflicts of interest should be disclosed to all stakeholders. It's then possible to decide upon the primary stakeholders and make sure all other stakeholders know where they stand in comparison.

Consider the work of a game reviewer – do you stand with the publishers as a cheerleader for games, or with your readers/viewers as an authentic voice of criticism? It's fine to be on the side of the game makers, provided you're not trying to convince your audience that they're the ones you're serving. As a software developer, are you interested in deploying robust feature-sets for your users or for ensuring the financial sustainability of your employer? What gets sacrificed when you pursue one over the other?

Resolving a conflict of interest requires deeply engaging with these questions.

Lazenby and McManus don't really have the luxury of saying that their external responsibilities are the ones with they are most dominantly aligned – their operational professional activities mean that they cannot recuse themselves from the day job. As such, their only real option was recusal and they did not exercise that option. Or at least, so this article would have us believe.

But under all of this, we see another ethical issue – a kind of debunking of the myth that research funding – and professional success in general – is a meritocracy. In truth, the quality and originality of a research proposal is only somewhat correlated with its eventual success (Cicchetti, 1991; Lane et al., 2022; Pier et al., 2018). The collaboration to which researchers are invited is largely a function of 'culture fit'. Those that are prohibited from alcohol by religious or personal convictions can find themselves locked out of social activities, which may leave them largely forgotten when it comes to promotions and preferment. On the other hand, a regularly shared cigarette break with a superior can have positive career implications simply by virtue of time spent together.

We can detect here in our case study the scent of cronyism where colleagues help each other navigate the treacherous and difficult terrain of research accomplishment. Funding rates of under 10% are not uncommon in the modern research economy – as in, for every ten grants submitted only one will be funded. That stark statistic hides the real truth though, which is that 10% gets a lot higher if you know the game and have a second player to help you along – something known in the literature as organisational proximity (Mom et al., 2021; Sandstrom & Hallsten, 2008; Travis & Collins, 1991). 'Networking' in many cases is a synonym for a systematised and ongoing conflict of interest that excludes many – particularly women (Schiffbaenker et al., 2022), people with disabilities (Swenor et al., 2020), and minorities (Nikaj et al., 2018) – from getting a fair hearing in supposedly 'objective' processes. John Blackbriar got his funding – who were the people that **didn't** get funded because McManus helped convince a funding panel that his university was the optimal destination for the money?

THE LONG MEMORY OF THE INTERNET

Our primary lens on this scandal is through the newspaper articles to which we are given access. This is an intentional design of the case study, to encourage a sense of critical scepticism about what we read. Or, perhaps more accurately, what we don't read. In the section above for example, we talk about Lazenby and McManus as if they hadn't disclosed their interests. Maybe they did and that fact has simply been omitted from the reporting in order to make it juicier. It used to be said that people turned to the news to find out if what they were reading on social media was true. Increasingly, people (including journalists themselves) turn to social media to find out if what they see on the news is true (Brandtzaeg et al., 2016; Zubiaga & Ji, 2014) albeit with poor results. Problematising this shift in behaviour though is that news and social media have become increasingly polarised (Barbera, 2020; Levy, 2021; Osmundsen et al., 2021). That newspapers pander to their audience is something that has been recognised for decades (Beam, 2003; Mullainathan & Shleifer, 2005), as is riffed upon within the sublime (if somewhat dated) sketch from *Yes Minister* (Kamm, 2016):

> **Hacker**: Don't tell me about the press. I know exactly who reads the papers. The Daily Mirror is read by people who think they run the country; The Guardian is read by people who think they ought to run the country; The Times is read by the people who actually do run the country; the Daily Mail is read by the wives of the people who run the country; the Financial Times is read by people who own the country; the Morning Star is read by people who think the country ought to be run by another country, and the Daily Telegraph is read by people who think it is.

In our latest article we see two effects at play. One is that citizen journalism – or student journalism in this case – has done important investigative work that would more traditionally be considered the role of newspapers or television news outlets. But also, the outlets that have now ceded responsibility for this kind of work have no compunction in then using it as the basis of their own coverage. There exists now a whole ecological niche of journalism where the job of the reporter is to find 'news stories' on Twitter, Facebook, or Reddit and then monetise them by directing attention to their own coverage (Hurcombe et al., 2021; Johnston, 2020). Sometimes with disastrous impact, such as the association of Sunil Tripathi with the Boston marathon bombing (Starbird et al., 2014). Indeed, whole outlets have this kind of content looting as the thing that drives the majority of their traffic – collated and 'curated' lists of Reddit posts and collections of tweets with a light smattering of original commentary to tie it all together (Tandoc, 2018). Often, these come along with a predisposition for audacity to hide the lack of originality – many of the original clickbait headlines evolved from a need to drive eyeballs to content those eyes had already seen elsewhere. Much of it amounts to little more than an ongoing creativity burglary. As Stehling et al. (2018) put it:

> We argue that processes of co-option constantly oscillate between the empowerment and exploitation of audiences.

Note for example how the Dunglen Chronicle refers to 'the Facebook poster revealing the information'. They do at least name the Facebook group (Professor not Patsy). Despite this revelation being worthy of a newspaper article, the poster is not worthy of accreditation. However, there is another edge to that particular sword – media attention can light the touchpaper under the firework that is social media. We can even see from the name of the group that this isn't a calm, reasonable place for people to discuss the situation. It's factionalised in its very framing, and undoubtedly at war with another Facebook group somewhere called 'Blackbriar is a Fraud'.

Where this kind of relationship between social media and the news can work well though is a kind of news version of Linus' Law (Raymond, 1999). 'Given enough eyeballs, all bugs are shallow' can also be recast as 'Given enough eyeballs, all information is accessible'. It's just not feasible for a small journalistic outlet to pore over the many thousands of terabytes of content that might be dimly relevant to a complex emerging news story (Cervi, 2019; Stehling et al., 2018). Citizen journalism often reflects a kind of crowd-sourcing of the kind of work a researcher might perform at a larger outlet. In this case, the unattributed original poster serves as the spark that sends the newspaper to the Freedom of Information Act (FOIA). We can tell this is a fictionalised case study by how the act enables them to get actual information back. Most FOIA requests end in a full or partial dismissal of the request[1] under any number of its commercial or sensitivity exemptions. In 2021, of over 51,000 FOIA requests only 40% were granted in full, and 38% were withheld in full. It's been argued that the FOIA in some cases has resulted in public bodies taking sensitive data entirely off-grid (McConville, 2021).

The grounds under which a FOIA request in the UK can be rejected are comprehensive. You can simply reject a request in its entirety on the ground that it would be too costly in time or money to process (27% of withheld requests), or that the request is vexatious or a repeat from the same person (4%). Of course, an external party critiquing a response on the first ground would find it difficult to prove either way, and vexatiousness depends on subjective criteria such as 'proportionality' or 'justification', or 'tone'.[2] It is the information holder, absent a legal challenge, who rules on these criteria.

And if this wasn't enough, provisions in the GDPR can also blanket-protect information holders from making certain revelations in specific categories. There some specific exemptions that can be employed, and these account for a staggering 69% of withheld FOIA requests. Some of these are information intended for future publication (practically anything in a university); research information (ditto); information relating to security bodies and national security; information related to investigations and prejudicial to law enforcement; public policy and the effective conduct of public affairs; communication with the royal family; personal information; confidentiality; trade secrets; and more.

Note that all of these exemptions come with guidance as to when they can fairly be employed, and some of them are because revelations are subject to other legislation (such as data access requests or environmental impact requests). But it's safe to say that a party looking to obscure information from a FOIA request has plenty of avenues to explore. These exemptions vary on a country-to-country basis, but legislation has a difficult job to do in balancing the public right to know and the organisation's right to effective internal governance. The wrong piece of data, absent any of its surrounding context, can do tremendous damage. While not obtained under the Freedom of Information Act, we can harken back to the Climategate controversy (Leiserowitz et al., 2013) in which the leaked emails and files from the Climatic Research Unit at the University of East Anglia were skewed and misrepresented to support a climate change denier agenda (Garud et al., 2014). An off-handed joke, or a careless use of a name, might be all it takes to stoke the fires of a conspiracy.

When there is an embarrassing revelation like this, the first thought of many bureaucrats is to stem the tide by closing off access to all currently accessible records. An audit will likely need to be performed to see just how bad it's going to get, and that's not helped by having a ticking time-bomb of public revelation as thousands of interested parties start poring through all the original records. Public bodies have a duty of transparency in many cases, and it is never obvious except in hindsight what transparency is going to be problematic in a scandal. Maybe in one case it's the date of a meeting. Maybe in another it's the list of members who give their apologies. In another it may be that a particular issue was discussed and never followed up on. The problem with information security is that data turns into information in unexpected ways and in unanticipated constellations. We've already discussed this in relation to, for example, how content-free metadata can provide the content it is missing by evaluating it in conjunction with other metadata. If someone phoned a doctor, then their partner, then an abortion clinic, we don't need the conversation log to know the topic of those conversations.

The bureaucrats at the EFC here have done what seems to be the initially obvious act of self-preservation – closing off access to public documents until the extent of the damage can be assessed. This doesn't necessarily imply a cover-up, but rather that it's best for a public response to be based on having more information about the emerging scandal than the public itself. It gives some distance from the immediate damage of revelation so that a cooler, calmer response can be constructed. It permits for the reputation of an organisation to be protected, which as we have already discussed is one of the primary tasks of an organisation. Prohibiting access to the files – in the short term – is the administrative equivalent of bandaging an open wound.

Remember 'The Internet interprets censorship as damage', as John Gilmore once said. Its systems for dealing with it are not robust or reliable, but they are situationally very embarrassing. Thus, the sensible Twitter user makes use of auto-delete features that expunge tweets older than a certain age. Nobody knows what the social media mavens of the future will consider to be a hellworthy trespass, and nobody knows when they will unexpectedly swim into prominence in online discourse. A full Reddit post history is a liability. Open access to your Facebook feed a disaster. And so it is here – a prominent public body would certainly find its content backed up on the Internet Archive, and it's the first place a certain type of internet denizen will check. In the attempt to bandage their own data wounds, the EFC in this scenario has just directed laser focus to the most vulnerable parts of their injury. The Streisand Effect once more comes to bear to excoriate anyone who wants to revise the past in line with the present's emergencies.

Again, it's important to stress here that the act of (presumably temporarily) closing off access to compromised files is not in itself evidence of malfeasance or a cover-up. It can be a sensible and necessary first step in auditing what has happened. News and social media are driven by novelty and by the virality of content. Mature, sensible reflection doesn't get retweets. Knee-jerk 'hot takes' is how you get yourself noticed and retweeted. And in that, it works best if you play to your factional interests and link up with fashionable grudges. Superficiality of insight is not an anchor on the speed of retweets – in fact, it might well be an accelerator as it inspires quote-tweets in defence or in dismissal. Such an environment is not conducive to a social media manager who is looking to provide robust, meaningful responses to a social media storm. It especially doesn't help cool, considered decision-making to have to constantly engage with the fringes of the discourse. It makes perfect sense to throw some cool water over speculation – to remove the kindling before it catches fire. The problem is though that social media fires burn hot enough to break cool water down into hydrogen and oxygen, and it turns out that instead of removing kindling you just replaced it with fuel of a more efficient nature.

NOTES

1. https://www.gov.uk/government/statistics/freedom-of-information-statistics-annual-2021/freedom-of-information-statistics-annual-2021-bulletin
2. https://www.legislation.gov.uk/ukpga/2000/36/contents

REFERENCES

Barbera, P. (2020). Social media, echo chambers, and political polarization. In *Social media and democracy: The state of the field, prospects for reform*. Cambridge University Press.

Beam, R. A. (2003). Content differences between daily newspapers with strong and weak market orientations. *Journalism & Mass Communication Quarterly*, *80*(2), 368–390.

Brandtzaeg, P. B., Lüders, M., Spangenberg, J., Rath-Wiggins, L., & Følstad, A. (2016). Emerging journalistic verification practices concerning social media. *Journalism Practice*, *10*(3), 323–342.

Cervi, L. (2019). Citizen journalism and user generated content in mainstream media. New dialogic form of communication, user-engagement technique or free labor exploitation? *Revista de Comunica, cao Dialogica*, (1), 120–141. https://www.e-publicacoes.uerj.br/rcd/issue/archive

Cho, C. H., Jung, J. H., Kwak, B., Lee, J., & Yoo, C.-Y. (2017). Professors on the board: Do they contribute to society outside the classroom? *Journal of Business Ethics*, *141*(2), 393–409.

Cicchetti, D. V. (1991). The reliability of peer review for manuscript and grant submissions: A cross-disciplinary investigation. *Behavioral and Brain Sciences*, *14*(1), 119–135.

Dallas, L. L. (2011). Short-termism, the financial crisis, and corporate governance. *Journal of Corporation Law, 37*, 265.

Felo, A. J. (2001). Ethics programs, board involvement, and potential conflicts of interest in corporate governance. *Journal of Business Ethics, 32*(3), 205–218.

Garud, R., Gehman, J., & Karunakaran, A. (2014). Boundaries, breaches, and bridges: The case of Climategate. *Research Policy, 43*(1), 60–73.

Hurcombe, E., Burgess, J., & Harrington, S. (2021). What's newsworthy about 'social news'? Characteristics and potential of an emerging genre. *Journalism, 22*(2), 378–394.

Jaffe, J. F. (1974). Special information and insider trading. *The Journal of Business, 47*(3), 410–428.

Johnston, L. (2020). Social news= journalism evolution? How the integration of UGC into newswork helps and hinders the role of the journalist. In *The future of journalism: Risks, threats and opportunities* (pp. 96–106). Routledge.

Kamm, J. (2016). Ignorant master, capable servants: The politics of yes minister and yes prime minister. In *British TV comedies* (pp. 114–135). Springer.

Lane, J. N., Teplitskiy, M., Gray, G., Ranu, H., Menietti, M., Guinan, E. C., & Lakhani, K. R. (2022). Conservatism gets funded? A field experiment on the role of negative information in novel project evaluation. *Management Science, 68*(6), 4478–4495.

Leiserowitz, A. A., Maibach, E. W., Roser-Renouf, C., Smith, N., & Dawson, E. (2013). Climategate, public opinion, and the loss of trust. *American Behavioral Scientist, 57*(6), 818–837.

Lemmens, T., & Freedman, B. (2000). Ethics review for sale? Conflict of interest and commercial research review boards. *The Milbank Quarterly, 78*(4), 547–584.

Levy, R. (2021). Social media, news consumption, and polarization: Evidence from a field experiment. *American Economic Review, 111*(3), 831–70.

McConville, B. (2021). Public or private? Freedom of information and the Scottish struggle for scrutiny of public bodies. In *Research handbook on information policy* (pp. 238–249). Edward Elgar Publishing.

Mecca, J. T., Gibson, C., Giorgini, V., Medeiros, K. E., Mumford, M. D., & Connelly, S. (2015). Researcher perspectives on conflicts of interest: A qualitative analysis of views from academia. *Science and Engineering Ethics, 21*(4), 843–855.

Mom, C., & Van den Besselaar, P. (2021). Do interests affect grant application success? The role of organizational proximity. https://doi.org/10.48550/arXiv.2206.03255

Moore, D. A., Tanlu, L., & Bazerman, M. H. (2010). Conflict of interest and the intrusion of bias. *Judgment and Decision Making, 5*(1), 37.

Mullainathan, S., & Shleifer, A. (2005). The market for news. *American Economic Review, 95*(4), 1031–1053.

Nikaj, S., Roychowdhury, D., Lund, P. K., Matthews, M., & Pearson, K. (2018). Examining trends in the diversity of the U.S. National Institutes of Health participating and funded workforce. *The FASEB Journal, 32*(12), 6410–6422.

Osmundsen, M., Bor, A., Vahlstrup, P. B., Bechmann, A., & Petersen, M. B. (2021). Partisan polarization is the primary psychological motivation behind political fake news sharing on Twitter. *American Political Science Review, 115*(3), 999–1015.

Pier, E. L., Brauer, M., Filut, A., Kaatz, A., Raclaw, J., Nathan, M. J., Ford, C. E., & Carnes, M. (2018). Low agreement among reviewers evaluating the same NIH grant applications. *Proceedings of the National Academy of Sciences, 115*(12), 2952–2957.

Raymond, E. (1999). The cathedral and the bazaar. *Knowledge, Technology & Policy, 12*(3), 23–49.

Sandstrom, U., & Hallsten, M. (2008). Persistent nepotism in peer-review. *SCIENTOMETRICS, 126*(2), 1863–1865.

Schiffbaenker, H., van den Besselaar, P., Holzinger, F., Mom, C., & Vinkenburg, C. (2022). Gender bias in peer review panels – "The elephant in the room". In *Inequalities and the paradigm of excellence in academia* (pp. 109–128). Routledge.

Starbird, K., Maddock, J., Orand, M., Achterman, P., & Mason, R. M. (2014). Rumors, false flags, and digital vigilantes: Misinformation on twitter after the 2013 Boston marathon bombing. In *IConference*. https://doi.org/10.9776/14308

Stehling, M., Vesnic-Alujevic, L., Jorge, A., & Maropo, L. (2018). The co-option of audience data and user-generated content: Empowerment and exploitation amidst algorithms, produsage and crowdsourcing. In *The future of audiences* (pp. 79–99). Springer.

Swenor, B. K., Munoz, B., & Meeks, L. M. (2020). A decade of decline: Grant funding for researchers with disabilities 2008 to 2018. *PLoS One, 15*(3), e0228686.

Tandoc, E. C. Jr (2018). Five ways BuzzFeed is preserving (or transforming) the journalistic field. *Journalism, 19*(2), 200–216.

Terry, S., et al. (2015). *The macro impact of short-termism*. Stanford Institute for Economic Policy Research.

Travis, G. D. L., & Collins, H. M. (1991). New light on old boys: Cognitive and institutional particularism in the peer review system. *Science, Technology, & Human Values, 16*(3), 322–341.

Wilson, T. D., & Brekke, N. (1994). Mental contamination and mental correction: Unwanted influences on judgments and evaluations. *Psychological Bulletin, 116*(1), 117.

Zubiaga, A., & Ji, H. (2014). Tweet, but verify: Epistemic study of information verification on Twitter. *Social Network Analysis and Mining, 4*(1), 1–12.

CHAPTER 16

Resignations All Around at the University of Dunglen

NEWSPAPER ARTICLE

An exclusive report by Jack McKracken for the Dunglen Chronicle

The University of Dunglen, unable to contain the criticism regarding Dr. Blackbriar and the university's senior management team, today released a series of press statements announcing four staff losses. The first of these was the principal of the university, Professor Sir David Tumblewood, who announced that he was resigning to take on the vice-chancellorship of the National University of Kumrani in the Middle East. His statement made no mention of the scandal, but university insiders say that 'he jumped before he was pushed, and so they gave him a golden parachute. It's a big one too – he'll be crushed underneath it'.

Dr. Blackbriar has taken early retirement, effective immediately. All internal investigations against him have been dropped, although civic proceedings initiated by ScotOil and other NSAE funding partners will still proceed. He will not be permitted to retain his professorial honorific, and so he reverts to the Dr. to which his postgraduate qualifications entitle him. Dr. Blackbriar's lawyers issued a statement that said, 'Dr. Blackbriar continues to assert his innocence in this matter, but recognises regardless of

DOI: 10.1201/9781003426172-16

the outcome that his career in academia is over'.

Our insider tells us 'The early retirement deal he signed was very generous, but it contains within it a clause that prohibits him from instituting legal action against the university. The problem for the university is that he knows where all kinds of skeletons are buried, and so it was worth it for them to pay him off'. Rumours suggest that the package given to Blackbriar also contains strict non-compete and non-disparagement clauses, although as is typical the man himself is tight-lipped on the topic.

Sir Gideon Lazenby has tendered his resignation and has been selected by his local parliamentary group to stand as MP for the Conservative and Unionist party in the next general election. 'He knows as well as anyone else that a Tory in Scotland has more chance of being hit by lightning than elected', said a close friend of the former university treasurer, 'But he's got a different trajectory planned – governmental advisor and strategy planner, and this candidacy is his way to get his foot in the door. Let's face it, even with these allegations hanging over him he's still more trustworthy than the majority of parliament'.

Finally, Professor Ian McManus has tendered his resignation to take on a full-time role at the European Funding Council as Director of Ethical Research. 'Seems ironic', said our Dunglen insider, 'But the EFC figure that his already considerable expertise is only enhanced by this incident. He always acted more like a bureaucrat than an academic anyway, so maybe this is for the best'. This move is seen as surprising by many in the sector, who cite the bitter aftertaste of Britain's contentious exit from the European Union. 'At one point having the UK in a research funding consortium was like lacing someone's coffee with cyanide', said one academic. 'I guess now we're just swallowing everything'.

Sharon McAlpine and James Duncan, the two PhD students who were suspended as a result of the investigation, have not been reinstated. The new acting principal of the University, Professor Helen Hackett, issued a stern rebuttal to Karan Chandra, saying in a letter 'The standard of research ethics to which we hold all parties at the university has been compromised significantly of late. Your clients are guilty of negligence or wilful blindness in the pursuit of their doctoral studies. As such, they will not be reinstated by this university and we will aggressively protect ourselves against legal challenges to that decision. They were half way through their studies, so the remaining funding will be used to take on a new PhD student'.

Professor Hackett spoke to the Dunglen Chronicle earlier today. 'We are hopeful that the recent announcements regarding university staffing are sufficient to draw a line under this sorry incident. The university's reputation, collaborations and financial security have all been negatively impacted by this sad affair. We will now work within and outwith this institute to repair the damage that has been caused'.

In related and unexpected news, ScotOil today announced that an oil-rig scheduled for being dismantled next year will be given a reprieve. The Blackbriar Algorithm had identified that region of the North Sea as being prime for sustainable exploitation of deep, previously inaccessible reserves. 'We had given up hope, honestly', said a ScotOil insider. 'And then boom – turned out that the algorithm was right all along, we just weren't picking up the soundings we needed. We adjusted our equipment, changed our methodology, and there it was. There's still half a dozen sites though where we're running speculative rig life extensions that haven't turned up what we were promised. Maybe they'll end up showing the same results, but I doubt it. I guess the system works, but not reliably enough for us to base future plans on'.

The NSAE project has announced it will retire the use of the Blackbriar Algorithm until it can be tested further. While their press statement remains hopeful about the future, academics affiliated with the project are more downbeat. 'The whole thing only worked, theoretically at least, because of Blackbriar's research. Now we've got one end of the process working, the other end of the process working, and nothing to link them together. We've got phase one and phase three, but phase two remains nothing more than a question mark. I can't see the project lasting more than a few months on that basis, and then all the funding will be pulled. After all, the only person who really knew how it all fit together was Blackbriar. Even if the underlying system is salvageable, I hear Blackbriar can't work on it even if he wanted'.

These high-profile casualties of the BrokenBriar Affair are only part of the whole picture. Many members of staff no longer wish to be associated with a university with such prominent and negative international attention. Many have quietly handed in their notices and indicated their intent to obtain employment in other institutions.

'We've basically been asset stripped', said our Dunglen insider. 'Anyone who had a reputation worth protecting has done the only thing they really could – they've moved on to other universities. That way they get a little stain on their CVs but nothing that'll stick forever. If asked about it they can even say they took a moral stand – refused to continue an association with a university like ours. All we're left with now are the dross that couldn't get out. We haven't lost everyone of course – some people are tied to the area as a result of mortgages and such. The fire has gone out of them though – a lot now are just counting the days until retirement'.

He adds, 'We put in all that hard work, all that effort – we got so close to being a top twenty institution, but we failed. The only trajectory we have now is down – down the league tables, down the research assessments, and down the toilet. We'll be lucky if we avoid becoming bottom twenty, really. Previously we got budding stars like Duncan and McAlpine applying for our PhD programs. The new guy who's replacing them is called Stan Templemore – a nice fellow, but it seems unlikely he'll be quite up to snuff, frankly'.

Professor Hackett is more optimistic about the future. 'It's true we've lost a number of very gifted staff members, but that's the way of life in academia. People move on. They move away. Sometimes though – they move back. We'll suffer in the short term, that's certainly true – but we managed to climb our way towards the stars once. We can do it again'.

STIFLING DISSENT

And so we arrive at the finale. It likely could not have unfolded in any other way.

We're probably unlikely to hear anything directly from Blackbriar at this point – aside from his (alleged) social media outburst he hasn't really offered his own perspective anywhere with regard to what has happened. An insider, speaking to the Dunglen Chronicle, says that he's been given a generous payoff but also that he has agreed to refrain from pursuing future legal claims against the university. He's also seemingly signed on to strict non-compete and non-disparagement clauses.

Let's begin with the first of these. A non-compete clause is one which prohibits an employee from working with a competitor for a specific duration after termination of employment. Non-compete clauses are comparatively rare in academia. Movement between institutions is generally seen as a good thing, allowing for a cross-pollination of perspectives and for best practice to be promulgated between different universities. The allegiance of an academic is often only somewhat to their employer – many will assert a deeper loyalty to their discipline than to their university. Fullwood et al. (2013) report when talking of UK academics:

> They have a relatively low level of affiliation to their university, perceptions of a high level of autonomy, coupled with a high level of affiliation to their discipline.

What would a non-compete clause even mean in a sector like this? It is, after all, virtually impossible for an institution to lay an ownership claim to the contents of someone's head. Many contracts will stipulate that teaching materials (and sometimes ancillary materials related to teaching) belong under the category of work-for-hire. That is to say, they are owned by the institution that employed the person creating them. The knowledge that was the source of the material in itself is transferable. Sometimes you might have to leave your lecture slides behind, but you can always recreate their contents. Generally though universities like to keep ex-faculty on good terms – after all, today's departing colleague is tomorrow's collaborator in a multi-institution research consortium. The collegiate nature of academia is such that most within it feel that collaboration is the key to meaningful success. Academics serve as external examiners for other institutions. They work cross-departmentally and internationally on PhD supervisions. Most see themselves as being part of a broader and noble tradition of knowledge sharing. The idea of a non-compete clause is almost incoherent in that frame.

Non-compete clauses are much more common in industries where companies are in zero-sum competition over the best talent – which is to say, the gain of an employee by one company results in the equivalent loss for another. That's not true in academia – network effects are powerful and all benefit from a broad range of connections. Talent in technical fields, experienced managers, and executives – all of these often serve as custodians of commercially sensitive information that could result in direct benefit for a competitor. Companies don't want a CEO moving to their main rival and taking all knowledge of future plans and products with them. As such, a non-complete clause is designed to offer protection against that kind of dangerous trading in commercially sensitive awareness. Non-compete clauses are usually time limited, often limited in terms of employment area, and of varying enforceability depending on where in the world you live. It's a matter of course for employment contracts in the United States to include these – taking a job means agreeing in advance to the non-compete provisions. Around 18% of workers in America are under some form of non-compete constraints.[1]

Blackbriar's case is slightly different, since he has signed on to non-compete not within his employment contract but as a separate settlement with the university related to the termination of his position. In the UK, these provisions fall under the general category of 'restrictive covenants' and are often applied in the same frame as 'non-dealing' clauses.

Those work similarly but put constraints around the specific named parties to which the covenant belongs – as in, 'you can't work for ScotOil for five years'. While controversial, they are legally enforceable provided an employer can demonstrate that such clauses are necessary to protect themselves from reasonable risk. Whether it's legally enforceable in the Blackbriar's case is simultaneously a matter for courts to decide, and almost entirely irrelevant. It's hard to imagine what university will be keen to employ a disgraced professor within the time-frame between his termination and his retirement, and it seems likely that he has disgruntled his industrial partners sufficiently that he will find few friends there.

The non-disparagement clause on the other hand will make Blackbriar's settlement conditional on his refusing to say anything negative about the University of Dunglen, and likely also any significant person or personages associated with the institution. In other words, he can't release a tell-all autobiography if he still wants to be in compliance with his settlement package. We're once again not privy to the exact wording employed, but it is likely under the circumstances to be broad – applying to oral and written forms, as well as social media, video, and anything else that comes to mind. It may be time limited, but more commonly a non-disparagement clause extends in perpetuity. Anything Blackbriar has to say in his defence in the future will have to be filtered through this contractual limitation on his freedom of speech. We have to hope, for his sake, that his settlement package was worth the gag – academics tend to chafe against restrictions of this sort.

Why is the university doing so much to stifle possible dissent in the future given that the damage has presumably already been done? Perhaps because Blackbriar may know where the bodies are buried – acting as he did at the highest levels in the university, with regular contact with senior decision makers, he undoubtedly has plenty of gossip as to the inner workings of the institution. Some of that may show the university in an unfavourable light if revealed by an uncharitable mind. Blackbriar has little reason to think well of the university, one must assume.

However, we also need to take into account the suggestive tone of later reporting that talks about the algorithm that gradually became less of a focus as the newspaper dug deeper into the people around it. It turns out that the system does work – perhaps not reliably enough for what is needed, but there are implications that given further testing the tool might still deliver on what was promised. That is, after all, what Blackbriar seemed to be claiming in his discussions with senior management. The spanner in that process though is Blackbriar himself.

In 1999, the Mars Climate Observer smashed into the surface of Mars. It had cost $125M to build and send it, and it took ten months for its trajectory to bring into abrupt contact with the red planet. The failure of the project triggered an internal inquiry. It turned out the problem was pretty simple – not a matter of the complexities of orbital dynamics, but rather of non-discussed assumption. Different parts of the intricate system that controlled the spacecraft were developed by different teams. Those at the Jet Propulsion Lab were using the metric system for calculations. Those at Lockheed Martin worked in imperial units (Lamb, 2022). Each part of the system worked perfectly as a self-contained unit. Both teams did robust work on ensuring that was the case. In software engineering we refer to that as 'unit testing' – checking the systems expected outputs against inputs to make sure that actual results match what we want to see. Checking to see if independent units work together is known as 'integration testing', and if that had been done properly the mismatch would have been identified before it resulted in an embarrassing PR disaster for NASA. That though would have robbed decades worth of educators of a truly wonderful example in the importance of robust quality assurance regimes.

The point of that digression is this – Blackbriar is identified as the only one who could realistically handle integration of the systems sitting at each end of the larger process driving the NSAE Project. That's what his algorithm was actually for – acting as middleware that used soundings from sensors to direct drones to areas of interest. The drones likely work correctly. The soundings – with appropriate sanitisation – likely work correctly. Without a correctly working algorithm from Blackbriar, there's no way for one to influence the other. Blackbriar leaving Dunglen under a non-compete clause means that he won't be able to go to ScotOil or any of the other partners. There's no cushy consultancy gig awaiting him – the algorithm needs Blackbriar's understanding of the big picture, and Blackbriar is legally prevented from applying his wisdom to the situation.

The university, having done everything it can to distance itself from Blackbriar, has one lingering big threat on the horizon. What if they're **wrong**?

What if, in the end, Blackbriar actually got the system to work correctly and senior managers ended up cutting the University of Dunglen out of a share of billions in revenue? There is a certain kind of managerial mindset that thinks 'if we can't have it, nobody can have it'. The risk of reputational damage – of being seen as someone that cannot be trusted to make sound judgements – is of great importance to the senior managers at Dunglen as they navigate their exit from this already compromising situation.

We can say here that the whole project's management has to be called into question because of the centrality of Blackbriar to the success of the endeavour. Many of the issues we have seen are down to him occupying a critical role that no one else could fill. In formal management theory, this is sometimes called the 'key person risk', but software developers have a different term for it – **bus (or truck) factor** (Ferreira et al., 2016).

Specifically, 'how many people need to be hit by a bus before the project cannot be advanced'. For a long time, the bus factor associated with the Linux operating system was one – if Linus Torvalds was hit by a bus, the project couldn't continue. When identifying a low bus factor, it is incumbent on managers to raise it to an acceptable value. That value differs from project to project based on logistical and economic considerations. Essentially though no one in a well-managed project should be irreplaceable. Nobody should be so core that they become 'the standard you walk past'. The bus factor associated with Blackbriar, and the financial implications associated with any bus that careened towards him, argues that this situation should never have arisen in the first place. Someone should have noticed that he was irreplaceable and set in place measures for contingency planning. Blackbriar likely wouldn't have been happy, but is that a high enough cost when a whole industry is compromised as a result of placating his ego?

THE GOLDEN PARACHUTE

So, we reach the final stage of the scandal – the laying of blame. One thing we hope that these extended commentaries have done is show that blame is a difficult thing to fairly apportion. Everyone has a responsibility for what happened to a greater or lesser extent. Even in what seem like clear-cut ethical violations (Lazenby voting on research proposals that benefit his institution as an example of that) are unlikely to be so simple when considered at the time. And in those circumstances where the ethical path is followed, it doesn't mean malfeasance can't sneak in. A word from the right person in the right ear can distort proceedings whether you recuse yourself or not. We can never peer into the heads of our protagonists to assess their intention.

However, the Dunglen Chronicle is not required to take a fully measured and thoughtful approach to the narrative it has constructed over these articles. It just needs to bring the story to a close, and nothing is more effective at drawing a line under a scandal than a string of resignations. The toll here is significant:

1. Sir David Tumblewood, the principal, resigned and took up a vice-chancellorship at a university in the Middle East, comfortably away from the fallout surrounding Dunglen.
2. Dr John Blackbriar has taken early retirement. You'll see that he no longer has the Professor honorific. We'll come back to that.
3. Sir Gideon Lazenby tendered his resignation, and looks like he'll be standing for Parliament. Success is not expected, so one has to assume this is part of a longer process of rehabilitation.
4. Professor Ian McManus has given up his role at the university to take on a promoted post in the European Funding Council.

Those are the hunting trophies collected by the media, but they're not the only casualties. Sharon McAlpine and James Duncan, the PhD students, have been suspended. They don't have a track record of accomplishments to shield themselves from the consequences of what happened. Their hopes of attaining PhDs are crushed – their funding has been revoked and it is unlikely they can take the work they have done to another university. Who'd want this tainted research, or would take a chance on these tainted researchers? At best they'll need to start again somewhere else. More likely, they have been permanently shut out of academia.

But note here what's happened – those at the top of the university simply 'moved on', and everyone else was moved out. And the level of punishment received is arguably disproportionate to the level of moral culpability. How can that possibly be fair?

One of the key moral lessons we all learn at some point is 'Life isn't fair'. The parents of the authors of this book drilled this into us like a mantra from a young age. When we encounter injustice, it is our natural human instinct to be angry on behalf of the victims (Batson et al., 2007; Pillutla & Murnighan, 1996). We live in a world though where the underdog rarely wins. The rich prosper over the poor. Hard-work doesn't guarantee recognition or reward. And, perhaps most infuriating, it's much easier to prosper unethically than it is ethically. An old video game called Alter Ego gave the first author a moral epiphany when he was very young (Heron & Belford, 2020). It presents you with a scenario where you are hungry and you can eat your sibling's ice-cream sandwich from the fridge, or choose restraint. When you choose restraint, the game asks you how you feel about the fact that you'll never be rewarded for the restraint? It asks the question, 'How do you feel when doing the right thing is simply expected, and not rewarded?'.

Thus the central difficulty when it comes to ethical decision-making. There is no moral desert where people get what they deserve (Kagan, 2017). Good deeds don't necessarily get rewarded and bad deeds don't necessarily go punished (Hsieh, 2000). And when rewards and punishments are apportioned, we still find 'life isn't fair'.

Professor Blackbriar is now Dr. Blackbriar. Professor isn't necessarily a 'for life' title. It doesn't stem from the achievement of a specific academic qualification (such as a doctorate). It is a position that implies a particular staff grading, and that grading is dependant on entitlement within a university system. When one moves on from a university, so too does their Professor title – although as with many things, this is honoured more in the breach than in the observance. When one retires from a university, it is common for the university to confer 'Emeritus' status – that the title is retained as an honourary endowment. Blackbriar has retired – he is no longer 'allowed' to use the title Professor. He will not be getting an emeritus professorship, given the circumstances. However, the strength of this kind of prohibitions varies from country to country. Professor is a legally protected title in Canada, but not in Sweden, the US, or the UK. As such, the only cost that comes from Blackbriar continuing to refer to himself as Professor is that the university will deny it if asked.

The payment for this is that he received a 'generous settlement'. He's now retired, free to take on whatever responsibilities he wants (within the confines of his non-compete agreement), presumably with few financial difficulties. His punishment is in reputation alone, although perhaps the ominous cloud of legal action will hang over him for some time. For some people, that's likely to be bad enough. It's hard to imagine though anyone really remembering this incident in future years, and for those that do remember to recall its specifics.

The principal, undoubtedly cashing in on the positive exchange rate of UK academic prestige, has moved far away from the University of Dunglen to a country where he may find himself in far more financially agreeable circumstances. It's not clear where he's gone, but many middle-eastern countries recruit UK academics under contracts that are astonishingly generous in comparison to what they'd get at home. Tax-free salaries; paid trips to their country of origin; subsidised housing; bonuses for publications; generous discretionary research budgets. He may find it a little warm in comparison to the misty cool rain of Scotland, but he probably won't worry too much about that in his airconditioned office. Some may see it as academic exile, but it's likely an exile that leaves him many thousands of pounds a month better off. Elba, it is not.

Ian McManus steps up to a promoted post in the EFC – no longer having to worry about receiving funding, he now has a pivotal role in awarding it. One might expect the EFC to be reluctant to take him on in a more involved post, but one feature of senior management is that exposure to failure can be seen as 'excellent experience'. If you drop a piece of industrial equipment and cost your employer 100k, you're unlikely to be promoted on the grounds of 'Well, they're not going to drop one again'. On the other hand, those in the C-Suite can convert – through a kind of reputational alchemy – failure into promotion. The Peter Principle (Peter & Hull, 1969) argues that people in an organisation rise to their level of incompetence. In some scenarios though, incompetence is just evidence of incomparable training.

Finally, Sir Gideon Lazenby is standing as an MP in a constituency where he is doomed to failure. An odd end to a glittering career perhaps, but in many cases such activities are the stepping stone to greater things. Serving a couple of years in a 'service' position is the key that unlocks many future doors in policy think-tanks and advisory positions. Having 'failed MP' on your resume is not as much of a sore point as it might seem. Consider Britain's Nigel Farage, who failed seven times to be elected and yet still managed to be an influential figure driving the UK towards the cliff-edge of Brexit. Again, it is a sign of experience – an invaluable and necessary learning experience on the way to greater public or civic influence. Being a candidate for one of the larger parties in the UK is a subsidised way of raising your profile, of shaping the narrative of your rehabilitation campaign, and of getting your face on the television. It is risky for a compromised candidate to stand, but one of the benefits of university 'court politics' is that to the outsider the stakes seem impossibly low and the intricacies impossibly arcane. It's difficult to craft an attack ad around anything less stark than direct embezzlement.

But what of the real victims in all of this? We are predisposed to find villains and heroes in a story because humans are natural story tellers (Kent, 2015), in part because stories are unparalleled ways to make sense of a complex world (Bietti et al., 2019). What gets missed in such personality-driven narratives though is the larger social context in which these characters (real or imagined) function. Harold Zinn's revolutionary 'People's History of the United States' (Zinn, 2015) addressed this by centring its view of American history within the structure of the working class. It focused on people within the labour movement as they shaped **our** conception of the country's conception of **itself**. But even in this, sociological studies tend to favour those people who the data most effectively records. Mostly, this is the men who record the data (Perez, 2019). The real victims of our scandal are likely to be similarly invisible.

A final summation of impact then has to focus on the staff and students who are now going to bear the largest impact of the scandal – those that in future years may think of their involvement with the university as a kind of 'resume stain' that they cannot excise from their professional record. They may have had nothing to do with the scandal. They may not even know any of the people involved. And yet they will bear the brunt of the long-term impact because after the facts are lost to time, people will remember 'Dunglen – there was a big scandal there wasn't there?'. Every job application from this point onwards is likely to include, at the very least, some meaningfully raised eyebrows that need to be lowered during the interview. The Scandal in Academia is individual malfeasance with collective punishment.

And so the real cost – an almost instantaneous brain drain of the best and brightest. They'll want to leave as soon as they can, even if only to forge the narrative that they left for moral reasons. They're the lucky ones. Talented academics who aren't talented **enough** for their reputation to outshine their association with a tainted university – all trapped in a declining institution. These include parents who have to take into account their children's schools and friends, their mortgages, and their familial obligations. They may want to move. They may even have the prestige to go somewhere else. They may though not be able to rip their families out of an established comfort. And for those that don't have options – academia is an intensely competitive employment context – it's hard to imagine their morale is going to be very high for a long time. The university has been largely beheaded, and that carries with it an expectation of administrative turmoil. And then there's the impact on recruitment – Stan Templemore, the guy who missed out on the PhD early on in the scandal, is now the occupant of the PhD studentship carved out from remaining funds. It is the PhD students that are the beating heart of any vibrant academic community – they are the largely unappreciated bedrock of university research in any high-ranking institution. Having world leaders in their discipline available as supervisors is a draw for the brightest talent, and a generator of the research funding that pays for PhD studentships in the first place. That is going to be lacking. In the battle for the scant few jobs that academia needs to be filled from its gross overproduction of doctoral candidates, Sam Templemore and his contemporaries may already be among the casualties.

Professor Hackett, the acting principal of the university, takes a more positive tone but it's hard to believe this is anything other than Pollyannaish wishful thinking. The battle for league places in the UK is furious – we've already talked about that earlier in the book. It takes a lot of work and goodwill to meaningfully shift an institution up the rankings in a way that is anything other than luck or happenstance. The University of Dunglen will have precious little of that to call on in the future.

That brings us then to the conclusion of our section on blame and punishment. Everyone is to blame, because ethical failures are almost always bound to a social context. Nobody is to blame, because no one person created the circumstances or could have prevented the circumstances. Everyone has to deal with the fallout, because an organisation is about its people more than its policies and its structures. The baddies win. The goodies lose. The rich get richer. The wheel keeps turning.

It's not a very cheerful conclusion, but nowhere did we promise you a happy ending.

NOTE

1 https://www.washingtonpost.com/business/2018/10/18/even-janitors-have-noncompetes-now-nobody-is-safe/

REFERENCES

Batson, C. D., Kennedy, C. L., Nord, L.-A., Stocks, E., Fleming, D. A., Marzette, C. M., Lishner, D. A., Hayes, R. E., Kolchinsky, L. M., & Zerger, T. (2007). Anger at unfairness: Is it moral outrage? *European Journal of Social Psychology, 37*(6), 1272–1285.

Bietti, L. M., Tilston, O., & Bangerter, A. (2019). Storytelling as adaptive collective sensemaking. *Topics in Cognitive Science, 11*(4), 710–732.

Ferreira, M., Avelino, G., Valente, M. T., & Ferreira, K. A. (2016). A comparative study of algorithms for estimating truck factor. In *2016 X Brazilian Symposium on Software Components, Architectures and Reuse (SBCARS)*, IEEE (pp. 91–100). Maringá, Brazil. https://doi.org/10.1109/SBCARS.2016.20

Fullwood, R., Rowley, J., & Delbridge, R. (2013). Knowledge sharing amongst academics in UK universities. *Journal of Knowledge Management, 17*(1), 123–136.

Heron, M. J., & Belford, P. H. (2020). Do you feel like a hero yet? *Journal of Games Criticism, 1*(2).

Hsieh, N.-H. (2000). Moral desert, fairness and legitimate expectations in the market. *Journal of Political Philosophy, 8*(1), 91–114.

Kagan, S. (2017). Equality and desert. In *Theories of justice* (pp. 439–456). Routledge.

Kent, M. L. (2015). The power of storytelling in public relations: Introducing the 20 master plots. *Public Relations Review, 41*(4), 480–489.

Lamb, H. (2022). Metrication: A matter of national identity. *Engineering & Technology, 17*(10), 26–29.

Perez, C. C. (2019). *Invisible women: Data bias in a world designed for men.* Abrams.

Peter, L. J., & Hull, R. (1969). *The Peter principle* (Vol. 4). Souvenir Press.

Pillutla, M. M., & Murnighan, J. K. (1996). Unfairness, anger, and spite: Emotional rejections of ultimatum offers. *Organizational Behavior and Human Decision Processes, 68*(3), 208–224.

Zinn, H. (2015). *A people's history of the united states: 1492-present.* Routledge.

CHAPTER 17

Postscripts

SHARON MCALPINE INTERVIEW

Interview conducted by Shannon Matthews, co-host of the 'Tales from the Chalkface' podcast

Tales from the Chalkface is an Edinburgh-based podcast that reports on 'the extraordinary experiences of ordinary people'. Its focus is on education at all levels of the system, from primary schools to PhDs. Its two hosts are Shannon Matthews, a former senior lecturer in industrial relations at the University of Alba; and Bill Menzies, who is headmaster at a private school in the Perthshire region. It has a modest following of around 5,000 monthly listeners. For the past few months, they have been running a series on people who shifted out of academic jobs and into other career paths. This week they are talking to Sharon McAlpine about her experiences at the University of Dunglen.

Interviewer: Finally today on the podcast, we have with us Sharon McAlpine. Five years ago, Sharon was a rising star in academia, until she became the victim of a scandal involving her superiors. Things looked bleak for a long time, but today she has bounced back and is here to tell us her story. Hello Sharon, and welcome to the podcast.

Sharon: Hi. Thank you, I'm happy to be here.

Interviewer: Sharon, for the listeners who haven't heard of you, can you explain a little bit about your career background?

Sharon: Sure. So, I was always interested in problem solving from a really young age, and I thought algebra was the best thing ever to do

DOI: 10.1201/9781003426172-17

at school. My relatives thought I'd probably become an engineer. But I did a four-year degree where you get to take a lot of different subjects, and I discovered that I was more interested in learning about how the planet was formed than in calculating the amount of materials required to build a bridge. There aren't too many job openings for geologists, so I did a Master's degree in Data Science.

Interviewer: Smart choice, given how the world has developed since then.

Sharon: For sure. I managed to get a 2-year postgraduate research assistant post straight from that. It involved a lot of data analysis and I found it really rewarding.

Unfortunately those fixed-term positions tend not to get renewed due to how they are funded, and I was geographically tied to that city because I'd got married shortly after completing my Master's degree. My then husband didn't want to move. He was pretty secure in a well-paying job, with good prospects. When my contract was in its final six months and it seemed clear that nothing was coming up locally, we decided to start trying for a baby. We were pretty lucky, and I had my son, Jacob, a year later. I applied for a few PhDs and eventually, when Jacob was 8 months old, I got a funded PhD position in Professor Blackbriar's lab.

Interviewer: Was it difficult to return to work when your baby was so young?

Sharon: Men do it all the time, when their babies are only weeks old. But yeah, it wasn't all that easy. Jacob was teething and wasn't sleeping well at nights. Thankfully the PhD hiring process is slow, so I didn't actually begin the post until he was a year old and I was able to get him into the subsidised university nursery.

Interviewer: How did you find the PhD work initially? Was it what you had hoped it would be?

Sharon: It was somewhat overwhelming at first. Professor Blackbriar is – or rather, was – a big name in this field. The lab was really impressive: full of expensive equipment, and a few postdoctoral researchers who were already becoming publishing superstars. They never seemed to leave the lab. Luckily James started at the same time as me and

Interviewer: This would be James Duncan, who was also implicated in the events?

Sharon: Yeah. He started at the same time so I wasn't the only person feeling like a fish out of water. We bonded over Blackbriar's lack of interest in us as people, coupled with his excessive demands on us as researchers. We were both kind of thrown in at the deep end and left to sink or swim. I clung in there, we both did, but I sacrificed a lot of time with my son, and my husband, in exchange for what I thought was my ticket into a top flight research career.

Interviewer: It sounds as if Blackbriar wasn't the mentor you were hoping he would be. Can you explain what your working relationship was like, and how that impacted on your enthusiasm towards your PhD studies?

Sharon: We could go weeks without getting a meeting with Blackbriar. He'd be out of the office at meetings with ScotOil executives, or on research visits, giving keynote speeches at conferences in exchange for an expenses paid trip to a conference abroad. But then he'd be emailing us, from Paris, at 8pm at night asking for the results of our latest experiment. The first couple of times he did that I waited until the following morning to reply. But at our next meeting he shouted at me about it, and told me that was unacceptable. When he needs something, he needs it immediately, not the next day.

Interviewer: Is that what you expected going in to the job?

Sharon: Not at all. I pointed out that I worked regular hours because that was when childcare was available. He rolled his eyes and said babies that age go to sleep really early, and I should keep checking my email notifications and deal with work when the baby was asleep. He asked what my priorities were, like he thought that there was no question that my work should be my overriding concern – certainly over and above my family. I work hard, but I'm efficient and I have other interests and relationships. I knew he and his wife were separated at the time and I wondered if his career had something to do with it.

Interviewer: They did reconcile shortly after he left the university, according to our producer. Perhaps once the pressure was off, he became a nicer person?

Sharon: Mmm. I can't really say anything about that. He was certainly demanding with regard to us. He expected James and I to do the vast majority of his teaching for him, for one thing. I'd done a bit of lab supervision whilst working as a research assistant, but he dumped an entire course on us, including all of the lectures. It's a postgrad module, but there were still 80 students on the course. We got no training, and it took up about 20 hours a week for the 12 weeks the course ran.

Interviewer: Twenty hours of contact time?

Sharon: No, that's factoring in the prep time, lectures, labs, office hours and marking. Even so he still expected us to keep up the progress on the research project. James and I found ourselves working 60 to 70 hours a week. My husband was not supportive. The creche closed at 5pm so I had to take Jacob home, feed him and my husband, put him to bed, then start prepping for the next day's classes. My husband would bath the baby and handle the bedtime routine a couple of nights a week, when I was really stressed. He made it clear that he resented it. He acted like he was doing me a massive favour rather than taking responsibility for his own son.

Interviewer: That sounds like he got the more fun stuff to do as well, as far as care went?

Sharon: I don't think he thought of any part as the fun stuff. I think he wanted a baby because it was the easiest way to get his mother off his back. Even so I was making good progress with my PhD research, and I kept telling myself that I just had to push through for the next few years and then I'd have my golden ticket. That is, a PhD from a world-renowned research centre, and a glowing reference from a world leading professor in the field. I'd be one of the lucky ones who would immediately get a permanent lectureship or research position and I'd be able to focus on my own research projects. I was hoping to get back to working 45-hour weeks and meeting up regularly with friends.

Interviewer: It sounds like you felt as if you were a serf, working your way to buying your freedom. Is that fair?

Sharon: That's one way of putting it. Being a PhD student can be great at the right institution with the right supervisor. I read a book once that said a PhD supervision can be as intense as a marriage – that you will both spend a huge amount of time woven into someone else's professional life for your mutual benefit. Anyone who read the [ed Dunglen Chronicle] newspaper articles at the time knows that we were seen as resources to be utilised as much as possible. There was no consideration for our mental health or work-life balance. It was all about extracting as much work out of us as possible, hopefully without us reaching the point of total burnout – not because they cared about our well-being, but because time off for a mental breakdown would hinder progress on the project.

Interviewer: That sounds like a toxic work situation. When did you realise it was so bad, and did you consider getting out?

Sharon: I guess it's a little bit like boiling a frog. At first you think it's just because everything is new, and the initial literature review will be intense but then things will ease off when you move on to the practical work. But then they slowly add more and more to your plate until the teaching starts, at which point an additional load gets piled on at once and suddenly you're up to 70 hours a week no matter how efficient you are. But by then I was 4 months into the PhD and didn't want to drop out. I was worried it would look like I wasn't cut out for academia, and I had no other suitable career paths. Doing a PhD closes a lot of doors outside of academia.

Interviewer: Can you explain for our listeners why that is?

Sharon: You're over-qualified, academically, for most jobs so they think of you as a flight risk. That you'll flit off somewhere better when the opportunity presents itself. But also, a PhD is very narrow – you become the world expert in something incredibly specific. If you want a good job, you're better spending the three or five years or whatever building out a useful skill-set. The skills needed for academic research are only somewhat valuable outside of academia, and even then only in some very competitive niches.

Interviewer: That makes sense.

Sharon: Yeah. Quitting one though closes a lot of doors inside of academia without opening others. I felt trapped. I also still hoped that it would get better. I'd stayed in touch with some of the postdocs and lecturing staff at the previous institution where I worked as an RA [ed – Research Assistant], and they seemed to have enjoyable and rewarding careers. And Blackbriar's job seemed perfect: getting to jet off places, be a minor celebrity, and get your underlings to do all the grunt work while you take a large share of the credit for it. Not that I wanted to do that, but I thought I could make things better for everyone once I had some of my own funding.

Interviewer: So there you were, gritting your teeth, doing the work of two people, and hanging out for your PhD award. But then it got snatched away from you. Did you have any inkling or suspicions about Blackbriar before that first article in the Dunglen Chronicle?

Sharon: I was genuinely as shocked as anybody. I don't usually buy that newspaper. Or any newspaper as I rarely have time to read them. It was James who emailed me the link to the online version of the article saying 'Urgent: Read this now!', then he came over to my workstation as soon as I'd had time to read it. I had no idea that Blackbriar was massaging the data in any way. To be fair, he was so aloof and hard to get time with that I had no idea what he was doing most of the time when he was actually in the lab. But he came across as competent and rather arrogant. I wouldn't think to question what he was doing, or the data he was giving me access to. It seemed fine to me – there were no obvious red flags that I had noticed.

Interviewer: Did James voice any suspicions? And do you think it is possible that maybe you subconsciously thought things were a little off but didn't want to question them?

Sharon: I can't speak for James, but he certainly never mentioned anything to me, other than complaining that they wouldn't make some application JAWS [ed – Job Access with Speech, a screen reader application] accessible so he couldn't get access to the full raw dataset. I just assumed it was frustration at the way they didn't make the appropriate compensations for his disability, and treated him like an inconvenience whenever he raised any issues. I didn't think that he had any suspicions about the integrity of the data.

Interviewer: You didn't think it was odd that you weren't given full access to the raw dataset either? They wouldn't need to make any adaptations to make it available for you, so did they have a plausible reason for that?

Sharon: I didn't really think about it to be honest. Blackbriar said it was something to do with proprietary datasets and that he only had permission to share them with a few select people who had that level of clearance. But he was also like that about the good filter coffee – it wasn't for people like us. I just assumed that as a peon I wasn't worth giving clearance to. I didn't think he was trying to hide anything from me. I just thought he couldn't be bothered to put in any effort to do anything to make the lives of his PhD students better, whether that was getting accessibility software working for James, getting permission for us to access the raw datasets to aid our research, or letting us use his expensive coffee machine.

Interviewer: Did you talk to him about any of that?

Sharon: No. I didn't feel comfortable challenging him. Remember – he had my future in his hands, and he was a petty man. I don't even remember sending that email James leaked to the press. Whenever I asked for anything he would just be really dismissive about it and act like I was wasting his time. If I really wanted something, like a new monitor, I would go through one of the postdocs. They had some budget allocated to them, and were more approachable. Blackbriar just walked about like he owned the place and gave the impression that if you wanted a good reference you'd jump to it whenever he asked for something. He'd put his name on all your publications, your job as his student was to produce those publications, stay out of his way and not bother him otherwise. We had no option but to trust him.

Interviewer: So, was that trust, or hope at least, destroyed when you read the scandal in academia newspaper article? Or did you still have confidence in him?

Sharon: Honestly, I wasn't sure. I wanted to believe that it was a storm in a teacup and that he would be cleared of the allegations. My success was tightly bound up in his, after all. Even now, I'm not sure what actually happened – the university cut us out of everything as soon as they could. But even though he was my primary supervisor, I didn't actually know him that well. I hoped the best of him. That could have been a combination of wishful thinking – because I knew how closely my reputation was linked to his – and a belief that a large institution full of highly educated people could not possibly have employed a cheat for 30 years. They would have realised there was something untoward about his research, right? Of course, now we know that they not only knew, they pressured him to do it.

Interviewer: What happened to you after the scandal broke?

Sharon: James and I were suspended the very next day. He wasn't in a union. I still was, since I'd never cancelled my membership from when I had the RA job. But they said they couldn't help us because everything that was happening was part of the official procedure, and all we could do was work within it.

Interviewer: How did that make you feel?

Sharon: Abandoned. And then later betrayed when it turned out the president of our local union branch was all buddy-buddy with the senior managers. Apparently they'd told him 'we've got robust evidence that damns them all', and he never asked to see it or queried its existence. He just took it on faith that we were a lost cause.

Interviewer: What was it like to be suspended?

Sharon: The suspension was on full stipend whilst the investigation was carried out, but it was really stressful. We felt as if we'd been thrown to the dogs. We'd done nothing wrong, but we were being scapegoated and discarded as a sacrifice to try and protect those higher up. We just felt completely powerless. I felt really angry, but also frustrated and depressed. This could ruin my career, and I couldn't see a way out. When Karan got in touch offering to represent us for free, it didn't take long to decide to accept the help to fight back. There was nothing to lose because the university had already implicitly made it clear that there was almost certainly no comeback from this. So we might as well fight it.

Interviewer: And yet you still lost your PhD post. Can you tell the listeners how events unfolded, and what happened to you?

Sharon: It was a real shock. But after Karan got involved I was hopeful that maybe I would get my PhD funding reinstated. At least until it became increasingly clear that Karan saw us as a stepping stone to national attention. He wanted to win, because that win would be good for his agency. He was more interested in building his public profile than our legal case. For me though – I wasn't allowed into the university grounds, let alone the lab. But I had a paper I had been in the process of writing, and of course with those long working hours I had a copy of it at home, along with some screenshots and graphs of the data. So during the early part of my suspension I kept working on that paper. Then things got worse. Those emails that leaked made me furious.

Interviewer: Which ones in particular?

Sharon: How dare they suggest that I would be a good person to hire because I'm pretty and would brighten up the office! What kind of 1950s garbage is that!? I'm a damn good researcher and it's like they overlooked that in favour of tokenism. Their comments about my son showed how regressive their views were. Oh no, what if she needs time off to look after her child, or, heaven forbid, gets pregnant again and take months off to look after another baby. It's like we're back in the 19th century and women have no possible role in life other than as wives and mothers – or possibly teachers or nurses or nuns if they can't catch a man.

Interviewer: How did you feel about that?

Sharon: It all made my blood boil. I missed my son's first steps. One of the nursery assistants took a video of him walking for me. His first words were Dada, even though I did the majority of the child-care at home. I co-authored two papers and was first author on a further two during that RA post, and was more qualified than that arrogant nephew of the VC. But nope, I was just seen as a pretty face and diversity hire who had better be convinced not to have more kids until she'd done her duty for the research project.

Interviewer: I can see how that would be upsetting. Those really were appalling emails. Just how lightly they all spoke about the four of you, as if you were subhuman.

Sharon: Exactly! They didn't really see us as fully human like them. Or, at least, not important enough to be worthy of respect. Then those emails got leaked which showed a few members of the senior management team, including people who were meant to be investigating the fraud allegations, were putting pressure on John to publish. They were threatening his funding, which was in turn threatening us. Then he got drunk and started posting public messages about knowing where the bodies were buried.

Interviewer: Uh, for legal reasons I have to…

Sharon: Yes, sorry. The alleged drunk messages. He said he was hacked, but who knows really. Certainly with the payoff he got, he must have known where the bodies were buried. I got nothing other than my reputation destroyed. The internet can be a cesspit, and for some reason people were deciding on my guilt on the basis of absolutely zero evidence. The memes they posted about me were horrific.

Interviewer: Was it just memes?

Sharon: No! Some evil sods who were never caught decided to use AI to superimpose my face in multiple porn videos! I would never let anyone film me doing that. And I don't visit those kinds of sites so I didn't find out about it until one of those influencers covering the story messaged me a clip to ask how I felt about it. I didn't know what to do. I was horrified. I couldn't watch it. My husband did, and he was absolutely furious. Of course he knew it wasn't me, but it had a negative impact on our relationship. For a day or two it was as though he couldn't even bring himself to look at me. With everything else going on our relationship was already under a lot of pressure. It was one of the most degrading things to know that people were watching this and imagining it was me, and there was nothing I could do about it. I felt completely violated. I moved into the baby's room and cut myself off from social media. Of course, I couldn't just block everything out as I had to be in contact with James and Karan, and I couldn't stop checking that newspaper. But the worst thing is that at one point my son will be at school and his friends will find those videos. I'll have to talk to him at some point about how mummy didn't do anything of those things the videos showed. That's a conversation I have been dreading for five years now, and will be for five more.

Interviewer: So all of this took a toll on your marriage?

Sharon: It was so hard. He didn't know how to be supportive. I wasn't even completely sure that he believed I was innocent. He doesn't work in academia and he doesn't understand the politics or power relationships involved. He was also pretty stressed about the fact that I had probably just wasted a couple of years of my life. PhD stipends are very low: not much above minimum wage – less really when you add up the hours you actually work. If I'd gone on a graduate training scheme to become an accountant or something we'd have had fewer money worries and I'd have been able to spend more time at home. It was tough for him to read all the vitriol about me on social media too. I blocked it after a while but he didn't. My mental health took a hit due it all, and I suppose I wasn't being a very good wife.

Interviewer: Were your other friends and family supportive?

Sharon: Online, yes. But they mostly lived far away. I'd moved away to go to university and never moved back home. So my parents were a couple of hundred miles away. I'd made some good friends during my first degree, but most of them moved away after graduation: that's just life. I thought I'd made some friends at Dunglen, but when the scandal broke they were reluctant to be seen with me, or even to communicate via email. I guess with the hacking that wasn't a surprise. None of them wanted to be tainted by association with us. James was my only real support. And even he was distracted. When we found out he was the hacker, I understood why he hadn't been more available.

Interviewer: So, tell us what happened in the end. What was the outcome of the investigation, and did they offer any sort of apology to you?

Sharon: The investigation wasn't really concluded by Dunglen. It was a student who found the stuff they were trying to cover up. Some of the investigators were trying to keep their names clear whilst being in it up to their necks. The VC was putting enormous pressure on Blackbriar to apply for the funding when he was unsure of the algorithm, Lazenby was on the ScotOil board and didn't recuse himself for a crucial funding vote, and McManus, the 'ethics guy' himself, also didn't recuse himself from the funding council decision when there was both a conflict of interest and he knew the proposal was fudged.

Interviewer: They all left Dunglen, didn't they?

Sharon: Yes, but they went off to cushy new jobs, or got handsome payoffs like Blackbriar. I don't think the figures were ever disclosed but I did hear rumours it was in the 'low middle seven figures'. James and I just got pushed out the door. No apologies, no golden handshake, not even a decent reference. I had to threaten to sue for constructive dismissal to get a reference from them. The way the powerful get treated, it makes it easier for them to do stuff like this without consequence.

Interviewer: It really doesn't seem fair. What was your initial plan after being terminated by Dunglen?

Sharon: Despite everything, I really wanted to stay in academia. I really liked being a researcher, and it was all I knew. I'm really good at it too. I applied for some PhD and RA posts at the other local university, and some in other cities that were just about commuting distance. But I just got rejection after rejection. Not a single interview. Usually I get interviews for about half of the positions I apply for. It felt like I'd been blackballed. The scandal was old news pretty quickly, but the internet has a long memory. I even tried leaving my PhD studies off my CV, but that was lose-lose. Either they recognised my name and said it was dishonest to omit things from my CV. Or they hadn't heard of me but thought I'd been a housewife for 3 years. Things move quickly in scientific research and a gap of 3 years is a death knell as your skills are seen as obsolete.

Interviewer: At what point did you realise that you needed to change careers?

Sharon: I was kind of in denial for around a year and a half. Things move slowly in academia, PhDs and postdoc positions are advertised because they have to be. Often there is an internal candidate that is going to get it unless an external is miles better than them. So I kept telling myself that was what was happening. But not even getting any interviews was a new thing. My search radius expanded as the rejections grew, even though I didn't know how my husband would handle it if I did get a job 200 miles away. Professor Blackbriar was such a giant in the field that even applying for jobs down south – everyone knew who Blackbriar was, and that was that. After eighteen months of constant rejection I started applying for non-graduate jobs in the local area.

Interviewer: Did things improve once you changed the types of job you were applying for?

Sharon: Yeah. It was still tough. I left the PhD off my CV, and used my married surname. I'd kept my maiden name for publishing reasons as I'd published four papers before my husband and I got married. A publication track record is really important in academia, so you don't want to lose that or have to explain it to an employer, and changing surnames can mess with your h index. I was applying to supermarkets, the local council, DSS [ed – Department of Social Services], architects offices, estate and letting agents, dental practices. Really anything I thought I could do. They were more understanding about the three-year CV gap, but were still worried that I was overqualified and would leave. After a few interviews I finally got a job offer from one of the estate agents.

Interviewer: How did that feel, working in a job that made no use of your studies?

Sharon: It felt strange at first. It took a bit of time for me to find my feet as the other staff weren't sure how to act around me, and they told me off for being a bit too intense. My ability to be hyper focused for several hours at a time made me a bit too productive and I was told to slack off a bit or I'd make them look bad. I guess that's only to be expected when your frame of comparison for what a job is includes a seventy-hour work week. Our manager did notice my work ethic, determination and attention to detail though, and decided to give me a shot as an agent rather than admin support. The market was booming and they were looking to expand.

Interviewer: That sounds like a very different career path to what you were used to. How did you find the transition?

Sharon: I wasn't sure I'd be good at the sales side of things. Research is mostly a solo or small group back office type affair with no customer service. But actually, you spend a lot of time selling what you're doing: presenting at conferences, writing journal articles where you sell what you've done in the best possible light, networking with other researchers in case you can collaborate, pitching project ideas, writing grant applications, persuading your supervisor to give you the juiciest projects to work on. All of that needs you to sell yourself and the work. So my soft skills were much better than I'd realised. And selling something concrete like a house – literally concrete, sometimes – was easier than selling the effectiveness of unfathomable computer algorithms. At first I got the properties that were harder to sell, because the prime ones went to the most experienced agents. But the market was so buoyant that I managed to sell most of them quickly. The commissions were good too. Bear in mind that I'd only been getting about 20K per year as a PhD student, and the research assistant post had paid about 26k a year. The admin post hadn't left me any worse off than as a PhD student.

As an agent the base pay is low: minimum wage basically, but with the commissions I was making between 6k and 10k per month for a couple of years. That really took a lot of the pressure off.

Interviewer: That sounds like a happy ending?

Sharon: Kind of. I spent three years earning way more than I did in academia, and the work was much easier. I did have to do viewings some evenings and weekends, but that meant I got some time off during the day occasionally and got to spend a bit more time with my son. I also knew the market would crash, so we tried to live on what I'd previously been bringing in, plus the cost of my son's childcare until he got his free nursery place. I had also started looking to find a way to hedge against the inevitable crash.

Interviewer: How so?

Sharon: You get early access to a lot of property deals working in an estate agent, and one family had a flat they wanted to sell quickly. Their aunt had lived in it for about 40 years since she'd been widowed, and it looked like she'd never redecorated. It needed a complete renovation, which they didn't want to do. I had enough saved up to have a deposit to buy the flat, and got mates' rates from some of the trades who do work for the letting side of our business. It took a few months, but I managed to flip it for a nice profit, which I then put into the next one. After a few of those I bought and renovated a bungalow, and made about 90k profit on that. I've got a couple of flats I rent out now.

Interviewer: That's quite a change from where you were expecting to be at this time in your career, I expect. From researcher to property mogul.

Sharon: Haha, not really. I've stopped flipping as the market seems to be softening. The commissions are getting a bit more sparse, and there could be a bit of a crash. I've got a war chest there though now, so I can buy another one when the market starts to recover, and the rent from the flats will help tide us through until things improve. I also documented the flips on Instagram and have a good following on there, and a few of my followers pay me for property coaching.

Interviewer: Wow, that all sounds really impressive. Are you maybe glad that things have turned out this way?

Sharon: Yes and no. Obviously I wish I'd never been linked with Blackbriar, and those months of people hating on me on social media and all the career uncertainty were horrendous. I still have nightmares about it occasionally. But I'm satisfied in my current career, I'm in a better financial position, and my work-life balance is better. I'm still bitter about how unjustly I was treated, and about the way Blackbriar and the other senior academics were parachuted out. But I've made a success of things from a much tougher starting point than them, and I'm really proud of what I've achieved. Sometimes I think that I should be using my skills for a more worthy cause, like solving the urgent problems the world is currently facing. But I'm helping people every day at work, and I get a lot more thanks and appreciation for that than I did as an overworked PhD student. I had a rough time, but I think in five years it will be obvious that this was the right path for me.

Interviewer: How does your husband feel about all of this?
Sharon: We divorced a couple of years ago. You'd need to ask him.
Interviewer: Sharon McAlpine, thank you so much for talking to us!
Sharon: Thanks for having me.

STAN TEMPLEMORE INTERVIEW

Interview conducted by Emily Murphy, junior reporter for the Midlands Mouth

The Midlands Mouth is a primarily online 'lifestyle magazine'. It is based in Birmingham and achieves around five million hits a month. The ad-revenue raised from this supports a small full-time staff of writers, and additional funding comes in from Patreon and other subscription-based services. Given the geographical region in which it is situated, it often reports on stories from all around England. One of their featured stories this month is an interview with Stan Templemore, who has recently taken up post at a university in the region. His interesting backstory – brushed with scandal but not tarred by it – made him an obvious target for Emily Murphy's attention

Interviewer: Some of you may remember the name Stan Templemore from the Blackbriar scandal five years ago. Let's make it clear from the start that Stan had nothing to do with anything that went on there. He'd been interviewed for a PhD post but passed over in favour of someone else. Email leaks though showed him being referred to in a very dismissive way by the interviewers. For a short while, it became a bit of a meme to call someone a Stan if they were seen as unexciting. You've all probably seen the meme template of Stan's student ID where – sorry Stan – he looks a little slack-jawed and gormless. People would replace the name on his ID card with a joke, and send it to their friends or Blether it at their enemies. For a while he was one of the most recognisable faces on the internet. We caught up with Stan to find out how it felt to be part of the fallout from something he had nothing to do with, and whether he's still known as a mediocre man.

Interviewer: So, Stan, it would be good if you could start by telling us a little bit about your career before the Dunglen incident.

Stan: Hi, thanks for giving me the opportunity to share my story. So few people care about the bit players in the thing. I did my degree at Dunglen, and really loved it. I was sold on the way that research was being made a priority there, and the university was going in the right direction up the league tables. My honours thesis supervisor managed to get some funding for a one-year project which followed on directly from the work I had done on my thesis, and so I stuck around. That was a lot of fun, but I never got around to publishing anything from it. I guess that might be part of the reason I didn't get seriously considered for one of the PhD posts that came up. I really wanted to stay at Dunglen, so I applied for the Master's program in Data Mining. I figured I could beef up my data analysis skills and strengthen my chances of getting a funded PhD after my Master's thesis.

Interviewer: How would you characterise the scandal for those that may not have known about it at the time?

Stan: Even now, do we know what happened? All I knew at the time, and all I know now, is that there were some allegations of research misconduct centred around Professor Blackbriar. I don't think we ever really found out in the end what he had done, if anything. It all just quietly got hushed up and everyone involved moved on. I met Blackbriar a few times and honestly I don't believe for a moment he would have done anything untoward. He always struck me as a sound bloke. Anyway – some research data allegedly got manipulated and then all hell broke loose.

Interviewer: What was your initial reaction when the scandal first broke in the Dunglen Chronicle?

Stan: My first feeling was of having dodged a bullet. If Professor Blackbriar actually had been fudging data and falsifying results, just think what would have happened to me if I'd been one of his students. As it was, I was doing well on the Data Mining course and if I couldn't get a funded PhD afterwards there were plenty of corporate jobs out there that I could apply for. It was a really hot ticket topic at the time.

Interviewer: How long did that feeling of relief last?

Stan: Only a couple of weeks. Then they started really tightening up on IT security, which made it harder to do my actual coursework. That was frustrating. The atmosphere also quickly changed at Dunglen, and everyone got super paranoid. I realised that the scandal might devalue my degree. I might not have had the misfortune to work directly with Blackbriar, but just having a degree from Dunglen might now be a liability. Or at least not considered as sound as they used to be. Then that hacker one of the PhD students involved in the scandal I think leaked that email about us to the press. Or forged the email, who knows. I don't know if they ever established that. But it basically said I was competent but completely forgettable. That's a hard thing to read about yourself in the local newspaper.

Interviewer: How did you respond to that?

Stan: I didn't get a chance to respond. Journalists weren't interested in that side of things. Nobody contacted me for comment. What had been said about the successful candidates was also pretty shocking, but they were part of the main story. I was just someone who'd been dismissed with little thought, and journalists didn't see it as anything of interest to them. I wrote a short blog post about it, but it only got a handful of reads. Less than a hundred. It only got one comment, which was 'Cheer up, you gormless prick'.

Interviewer: Do you have any thoughts on why you were described in that way by the university? Can you remember anything about your interview? And how do you think you performed as an RA?

Stan: I thought I'd done a really good interview! I spent ages preparing my presentation, bought a new suit, and practiced all the answers to typical PhD interview questions beforehand. Maybe I was a bit overprepared. I sung the praises of recent papers Blackbriar had published, and also had only good things to say about my thesis and RA supervisor. I thought my enthusiasm, hard work and preparation would give me a good chance of getting one of the two posts. I was a bit blindsided to be dismissed as 'competent but forgettable'. And that line about 'hard to pick out of a line-up of one' was brutal! It really knocked my self-confidence.

Interviewer: It was absolutely brutal, I agree. They did say that you were well-liked and helpful to colleagues though.

Stan: Yeah, that would have come from the reference from Professor Holmwood, my honours supervisor who got me the one-year RA position. He always tries to say positive things about people.

Interviewer: I'm sure he sticks to the truth though.

Stan: Yeah, unlike Blackbriar he never got accused of fudging anything. I was quite popular as an RA. It was a small research team and Holmwood, whilst not as big a name as Blackbriar, also wasn't as much of a slave driver. We all took a proper coffee break every morning, and I took charge of making the tea and coffee as the junior member of staff. I had everyone's preferences and cups memorised within a couple of days. I also organised a games night once a fortnight and some of us took the opportunity to clock off and relax. We became good friends, bonding over something other than just work.

Interviewer: That all sounds great, but not things that would help with your research track record.

Stan: Yeah, that might be part of the problem. I'm also not that great at coming up with original ideas. Or at least I don't think I am. I'm more of a follower than a leader. I got a reputation as being a good listener, and a great person to bounce ideas off. Several of the researchers would come and chat to me when they were stuck with something, and if I could help I would. For example, one of the researchers was having a problem with modelling the growth of fungi through soil. The simulation wasn't matching what was happening in the physical experiments. I didn't know what the issue was either: I haven't studied fungi. But I just kept asking what were probably pretty pointless questions, until suddenly they had an epiphany. There was some factor that they had forgotten to include in the computer model, or to which that they had given the wrong weighting. I'm not sure which. But the next afternoon they were all excited because suddenly the model was showing much greater accuracy. They got a paper out of that work, and their associated funding proposal was successful. He was on the verge of abandoning the work as fundamentally broken, so I felt really good that I had been instrumental in helping turn a nothing into a something. Not only that, something significant.

Interviewer: That sounds great. Did you get some credit for that?

Stan: He bought me lunch, which was nice. I did joke that I should get a co-author listing on the paper. But he said that wasn't possible because it wasn't 'Vancouver compliant', apparently. I'd just had a chat with him for a few hours. He was the one who'd done the work and had the breakthrough: I was just a useful sounding board, which is hard to quantify in a paper.

Interviewer: That must have been disappointing for you. Did you focus more on your own research after that?

Stan: I tried to. But people liked talking to me, and I was too amenable and agreeable to tell them that I was too busy and needed to concentrate on my own work. I somehow became the lab sounding board and problem solver, and probably about 30% of my time was spent helping others with their breakthroughs. I got a lot of goodwill and cake from that, but unfortunately nothing more concrete. They got papers and sometimes funding, and in one case a permanent faculty position. I'd trade the cake for a permanent post if I could. I mean – where do you put 'had a project defining conversation with a colleague' on a CV?

Interviewer: So, was there any quid pro quo? Did anyone let you bounce ideas off them, or help when you hit a roadblock with your project?

Stan: Not really. I mean, my supervisor did, but he was busy so that was one hour a week I got of his time. I would mention things at coffee break sometimes, and someone might say, 'Okay, I'll come and have a look at it after I get my next experiment completed'. But then they would never follow through on that, and I didn't want to nag them. So I just plodded along by myself.

Interviewer: It sounds like those relationships were a little parasitic.

Stan: I'm not sure that's a fair way to put it. They were all good people: just with a really narrow focus. They didn't see that what they were doing wasn't in my best interests – they just didn't really think of me at all except when they wanted to test something out on an intelligent bystander. I think that's how a lot of researchers are. There's so much pressure to perform and publish and bring in big funding grants to have any hope of permanency that you can't really blame them for a bit of tunnel vision.

Interviewer: Given your experience of being sidelined as an RA, and being dismissed as forgettable for the PhD post, and your thoughts on the devaluing of a qualification from Dunglen, what on earth made you decide to take the PhD post after the scandal?

Stan: That's a fair question. I didn't actually apply for the post again. They contacted me and said that I had been the reserve choice and, since the successful candidates had not worked out, they were now in a position to offer the post. I've heard of that happening to a few people, but a month or two after the rejection maybe – in one case it was after 4 months as the candidate started and then quit after two. But getting an offer like that after nearly two years is pretty rare I would think.

Interviewer: So why do you think they did that, and why did you take the offer?

Stan: It was the easiest thing to do, to be honest. For all of us. They couldn't take Dick [ed – David Tumblewood's nephew, who also applied for the position], because that'd just fire the starting pistol for a whole new chapter in the scandal. I was a safe option. I'd taken a knock to my self-confidence, and I didn't have anything lined up after my MSc. I hadn't really started applying for jobs in earnest, but I'd had a couple of knockbacks. At the sole interview I went to, one of the other candidates recognised my face when we were in the group part of the selection process. During the coffee break he brought up the meme about me, and the other guys and girls had a good laugh at my expense.

Interviewer: Not what you want preying on your mind in an interview, I'm guessing.

Stan: It really wasn't. The interviewer never brought the meme up but I was still paranoid that they were thinking about it. I was second guessing the reason for any subtle facial expression they made. I got a bit flustered, and they gave me the 'You performed well in the group session and tests, but we decided to go with another candidate who was a better fit for the role.' type of unhelpful feedback. I was sure I wasn't going to get a job with that meme still circulating, and I was pretty despondent about it. I hadn't done anything at all other than apply for the wrong role at the wrong time. Nobody would have seen that email otherwise. But I also wanted to stay in academia rather than go into industry, and I figured with a new professor leading the research lab, and three years for everything to blow over, I'd have security for a bit and a PhD that would be worth something at the end of it. So I accepted the position.

Interviewer: Did that work out for you?

Stan: Well, I'm now Dr Templemore, so yes. It was a shift in topic, no longer attached to the oil-rig stuff. It was in computational biology in marine environments, which was ideal for my data mining background. The new professor didn't have the track record of Blackbriar. But he was also honest, and appreciated hard work, and had time for his students. It took a lot of 60-hour weeks, and I still went over my funded period by around six months. That was hard financially, but I'd assumed I would run over as the majority of PhD students in the UK do. I scrimped and saved to cover a year of writing up and job searching after the funding ran out.

Interviewer: How was the job search for you, and what are you doing now?

Stan: The job search was tough. It's easier in computational biology than in some other fields – I'd hate to be a humanities PhD for example. But universities even in hard sciences still pump out three or four times as many PhDs as are needed for the available positions. There were scores of applicants for every role. I got some good teaching experience during my PhD and knew I would love to become a lecturer, but the only place I managed to get an interview was at one of the bottom of the table English universities that used to be a polytechnic. Even they didn't offer me the role. They said I was their second choice, and they'd keep my application on file in case anything suitable came up. Even so it was in the South East of England and the pay wouldn't have been enough to live comfortably there. I'd need a flatmate at the least to afford a decent standard of living, and I'm a bit wary of putting myself in that kind of position.

Interviewer: What then?

Stan: After several months of rejections, I managed to get a two-year postdoc position at a university in the West Midlands. That's the one I just started last month, and I'm assuming the reason you contacted me. It's a mid-ranking institution within commuting distance of Wolverhampton. The cost of living there makes it doable. It's not quite the research area I had hoped to go into though. It'll get me some needed experience. I've got a few publications now, and I'm going to focus on getting as many as possible in the first 18 months there so that I can hopefully get a lectureship or a longer postdoc position lined up before the contract ends. It seems a nice place to work so far, but I've learned my lesson about being too generous with my time, and I need to make sure I focus on what's best for my career.

Interviewer: I'm glad to hear things have worked out for you, at least for the medium term. Has anyone mentioned the Dunglen scandal there?

Stan: My immediate superior knew Blackbriar, so he and I have talked about it a bit. And I think he got in touch with you to talk to me. Really though, nobody cares much any more. You mention Dunglen and people say, 'Oh, I recognise the name but I don't remember anything else about it'. It's old news, and five years is a lifetime in terms of memes. I've grown a beard, as you can probably see, and I wear contacts now instead of glasses. People have a vague sense they know my face, but can't place it. If I'd stayed in Scotland it might have been harder. More people would remember. It's tough leaving my family and friends to move here. It's even harder having my girlfriend of two years living so far away. We'd hoped to move in together, but the promised postdoc position at Dunglen never materialised. This was the only offer in town. We're going to try and make it work long distance, but I know people struggle with that. Whatever happens, I need to give this job my best shot so that I can get more stability. I'll be in my late twenties when this contract ends. People say you should strive in your twenties and thirties, consolidate in your forties, and then nurture the next generation in your fifties and sixties. What with what happened to me, I'm a decade behind schedule.

Interviewer: Thanks for talking to us Stan, and best of luck with your new job. I hope everything works out for you.

Stan: Thank you. So do I.

JACK MCKRACKEN INTERVIEW

Interview conducted by Jafna Jamal, a new segment host on Wake-Up London

Wake-up London is a television programme hosted by Southeast Enterprises Media. It broadcasts between 6am and 9am every morning, primarily to the London region but it is available as a Freeview broadcast and via live streams on the internet. It is a very popular show, with a viewership of around 200,000 a day. While these numbers do not stack up against equivalent shows on the BBC, they are remarkable for a regional television programme. Part of their appeal is in the controversial nature of their unabashedly subjective coverage of public scandals. Decried by the metropolitan elite as 'rank populism', they position themselves as speaking truth to power on behalf of the working people of London. Jack McKracken is a regular correspondent for the channel, but today he is invited on as the subject, rather than host, of an interview.

Interviewer: Welcome back to Wake-Up London. The time is 8:30am. In this segment of our show we are talking to newspaper journalist Jack McKracken, formerly a star reporter at the London

Panopticon and regular contributor to this very program. He is now going it alone to become head of the independent think-tank 'The Centre for Responsible Journalism'. Jack, welcome to you.

Jack: Thank you for having me.

Interviewer: You're very welcome. Now Jack, tell us – why now? You've been with the Panopticon for five years, and with us for three. Both after you shot to national attention as a reporter for your local newspaper. In fact, if I'm not mistaken, it is exactly five years to the day since you wrote the final article about the University of Dunglen and moved to London as your star began to shine?

Jack: That's right.

Interviewer: You shot up the ranks – interviewed everyone from musicians to actors to the Prime Minister to the latest winner of Big Brother. You've been tipped for a while as the next editor-in-chief of the London Panopticon, which has become the most popular print journal in the country over the past two years. Things have never been better for you, professionally speaking. Seems like an odd time to make an exit.

Jack: And they say print is dying, right? No, it's a fair question. And I guess the answer is… it's time for a change.

Interviewer: That's not much of an answer.

Jack: No, it's not. But that's the core of it. It's time for a change in my career. A change in my professional practice. But also, a change in this industry. And perhaps unsurprisingly, it was the five-year anniversary of the Dunglen scandal that triggered it. There were some undergrads at the University of Alba doing their bachelor's project on the scandal and they contacted me. They asked if I had any material that perhaps hadn't seen the light of day, and if I wouldn't mind talking to them and sharing my insights and unpublished documents.

Interviewer: And you were willing?

Jack: Definitely. You know, it's always pissed me off how often bright, go-getting kids are dismissed by more experienced members of their industry when they're trying to get ahead. Part of all of our professional duties are to help educate the young with the wisdom of the old, right? And it had been years since I had looked at any of the material and I was pretty excited about it being made part of the academic record. Even if only as a bachelor's thesis no one will read in a library no one will visit. So yes, it's the responsibility of more experienced, more accomplished people like me to nurture the spark of greatness in the young.

Interviewer: And this insight resulted in you quitting your job?

Jack: Oh, no. That's just an aside. Anyway, I agreed to meet them for a few hours to talk over the reporting. I looked out all my old files – believe it or not, I still keep most of my files in cardboard boxes. No worries about anyone slipping a sneaky edit into a print-out. I started reading through the extended documents, transcripts, and my own notebooks. It differs from place to place, but journalists are generally advised – and in some cases legally mandated – to keep their notes for a period of time after the reporting. Someone might sue for libel or whatever, and you need the evidence to hand. I generally don't throw out any – I have an attic full of files.

Interviewer: Mrs McKracken must not like that.

Jack: It's another Mr McKracken actually, and he understands.

Interviewer: I see. Please, do go on.

Jack: Yeah. Anyway, I was looking through these files. The Dunglen piece was my first big – uh, nobody uses the word 'scoop' but I can't think of a better one. I was just out of university, trying to carve out a name for myself. I took a shi… uh, I took a rubbish job at the local paper. I was a 'junior assistant reporter', which was basically code for 'the kid that makes the tea run'. I was allowed to type up some reports and maybe get a few of my own articles in if they were small enough. 'Jam jar smashed at local farmer's market' kind of thing. I'd been doing that for maybe five years, slowing rising up the ranks until I was 'assistant reporter' and then 'reporter'. And there I languished. Then, out of nowhere, I get this anonymous email from someone. I was given Blackbriar's name, and some vague assertions regarding the quality of his research and the implications of misconduct. And then it was up to me to turn some attention on to it.

Interviewer: If you were so unknown, why you?

Jack: Good question. Perhaps they wanted someone pliable? Perhaps they thought I'd do the grunt work and then someone higher up would bigfoot me. Perhaps they wanted to rattle some cages but didn't want it to become an international story. I don't know, and the informant never contacted me again.

Interviewer: Do you think you were being used?

Jack: Yeah. Not well, but yeah.

Interviewer: Interesting. So, back to the files?

Jack: Yeah – I started auditing my own reporting, drawing on the years of experience I had gained since then. And honestly, I was surprised by what I saw. Surprised, and a little shamed.

Interviewer: Really? Why?

Jack: I just don't think I gave an objective sense of what was going on. I wasn't trying to craft a narrative, but it's obvious from email chains and memos that a narrative was being constructed through me.

Interviewer: What do you mean?

Jack: Okay, let's take a simple example – the leak of emails that Nemesis –

Interviewer: The fake ones?

Jack: They weren't all fake. Many – I would say most, even – were authenticated by secondary sources.

Interviewer: Oh, I didn't realise that.

Jack: No, right? You didn't realise it because we never really emphasized it. While we did say that several of the emails were authentic, the simple existence of fake emails completely corrupted that source, both for our reporting and in the minds of the general public. The work needed to authenticate emails was considerable – remember, at this point the university was starting to clam up and getting anyone to speak – on, or off the record – was increasingly difficult. So we stopped using the emails as the basis of stories because we couldn't corroborate – or disprove – any of them.

Interviewer: It seems like you had plenty to report on though without the emails – what was the problem?

Jack: The problem is that we'd only really started to make use of those emails and the ones we picked to begin with were the juiciest ones. By the time they ceased to have value for the reporting, the ones that were calmer or more complex simply didn't 'progress the story'. They were too nuanced, needed too much context, or made the chronology of events 'too confusing for the reader', as my editor put it.

Interviewer: So they never got reported?

Jack: No, and I think some of them would have contributed a lot of signal to the noise.

Interviewer: For example?

Jack: Blackbriar – sorry, the person that hacked Blackbriar's Blether account – said that McAlpine and Duncan knew the selection criteria for the data. That email chain isn't in the leak, but if you look closely at what is there you can see sort of – phantoms? – that the conversation took place. You'd see fragments of it quoted elsewhere, and people would refer to the discussion in relation to other things. In other words, it's a bit like gravitational lensing – you can infer the existence of the missing email from the way the rest of the emails bend around it. We should have dug more into that – as difficult as it would have been – because if someone is hiding something you know it's going to be good.

Interviewer: You're saying the email collection you were sent was censored?

Jack: Selectively edited, yeah. It's not like seeing that email would have exonerated Blackbriar but it certainly would have cast some events in a different light. I saw some quoted correspondence for example that suggested the girl – Sharon – was actually a bit of a troublemaker. Seemed like she expected the whole university to bend around her needs just because she had a kid. But the actual emails themselves – no sign of them.

Interviewer: But you didn't edit any of the emails yourself, right? This is how the information came to you? Why do you feel bad about not pursuing some leads when it was clear you had enough to occupy you? All reporting is biased, after all – we're the ones who have to choose which leads to pursue, and we can't pursue them all equally.

Jack: I think we made a big deal of what was clearly a tainted source, and when we discovered it was tainted we should have done some in-depth analysis to find what part of the story wasn't being told. I pitched that idea to my editor, in fact.

Interviewer: What did he say?

Jack: He didn't say 'no'. What he did instead was give me a piece of advice that has influenced my career since. 'You don't have enough time to do all the things you need to do. Which of the things you **need** to do are you going to sacrifice for this thing you **want** to do?'. The answer was 'none of them', so the feature went onto a 'to-do list' for when I had time. And then I just never had time.

Interviewer: If you'd had more time, you would have dug more deeply into the data?

Jack: Yeah, I think so. I mean, there's a whole chapter about it in my new book. Or if there were more people on the story we could have devoted more time to corroboration and research. In the end it was just me. I did ask if I could get an intern or something, but I got the impression that they were less interested in an authentic story than they were in the massive ad-revenue they were suddenly getting.

Interviewer: The story was successful, yes?

Jack: Of an order of magnitude beyond what we'd normally expect from a local paper. I mean, it had everything right? Secrets and lies. The powerful being held to account. Billions of pounds at risk. A photogenic female victim, alongside her blind friend. A pompous clown of a principal who made things worse every time he opened his mouth. In the end the story is almost entirely about where the boundaries between professional practice and malpractice lie, but you could miss that in all the drama. But yeah, we were selling ten times as many papers. And then the internet got hold of it.

Interviewer: What happened then?

Jack: Our servers broke. Suddenly we were getting hundreds of times as many hits as before. It was like our version of when Kim Kardashian showed her bum and broke the internet by causing web traffic to spike beyond what its architecture could sustain. Blackbriar had done the same thing – he'd 'shown his whole ass', as the kids on the internet say. It took hours to get back online, and when we did our views started to spin upwards like a bandsaw. A story at the Dunglen Chronicle might get ten or twenty thousand hits online. These stories got tens of millions, and even now they still get far more attention than the average article. I think the editorial policy was 'Don't ask, but tell anyway'. At least, that's how it was implicitly communicated to me.

Interviewer: So the story was sexed up for the views?

Jack: No! No, nothing like that. We didn't print an untrue statement. We didn't make an untrue claim. But we had to pick and choose from a lot of elements as to what would make the cut for an article and I don't think we always picked them in service of the truth. We picked them in service of the story. It wasn't intentional, you understand – we were just reflecting our own biases and leaning towards what was in our own best interests.

Interviewer: Do you have any other examples of where the coverage may have not reflected the whole truth of the matter?

Jack: There's a lot of debate in my notes about wording. We tried to maintain a journalistic framing throughout, but there are a few times where subdued headlines of mine got adjusted to be a bit more sensational.

Interviewer: For example?

Jack: One article that I wrote had the headline 'Question marks around potential multimillion-pound consequence for research project at Dunglen University'. Okay, fair enough – too wordy. But I didn't get consulted on it being changed to refer to a 'research fiddle'. For one thing, that's not in alignment with the text. For another, the headline is important because it anchors everything that follows. I don't think, at that point, we could realistically say that there was a research fiddle at all. No actual misconduct had been identified.

Interviewer: Isn't that what editing is supposed to do though? Take content and change it into a more appropriate form?

Jack: Yeah, but it shouldn't change the essence of the story in the process. Or the first impression people should get when they encounter it. That kind of thing happened a few times in the chain between me submitting my copy and it being printed.

Interviewer: What else?

Jack: I think I'm guilty of some selective interviewing in later phases of the reporting. When I went out to interview students, for example, I probably wouldn't spend as much time with those that wanted to put forward a nuanced view of what was going on. In my defence, it's hard to express complexity in the word count of a newspaper article. I wasn't writing think-pieces for the Guardian or anything. This was a local newspaper where a story could be trashed if it was going to take space away from the football results. But it's clear looking through my transcripts that the largest attention was given to people with the most outspoken views.

Interviewer: What was the effect of that, would you say?

Jack: The 'culture of fear' – another headline I didn't have a say in – was a bit overblown I think. Really it felt like any other campus on any other day – at least, until the influencers came along and added a tension. It was a self-fulfilling prophecy really. If we hadn't made the campus sound like it was in the shadow of a Soviet-era Kremlin, we probably wouldn't have seen so many online personalities descend on it. In turn it wouldn't have been a campus riven by tension at all. We didn't become the story, but we did become the reason the story went the way it did.

Interviewer: How useful was it to have so much online attention centred on the university?

Jack: Ha. On one hand – well, there are the interns I wanted. I could trawl through hours and hours of videos to find information I had missed, and it served as its own interview evidence. But I always felt a bit grubby about it – some of the tactics used to jack up the views of these videos were pretty far out of alignment with traditional mores of reporting. Still, if the information is there and auditable why not use it?

Interviewer: Did you provide proper attribution of information that came from online sources?

Jack: I did, in my copy, but it was often muted by the time it got to print. 'This linked video shows this and that' would become 'Online sources say'. That kind of thing. To be fair though, the internet is a wild-west of plagiarism and copyright infringement. It's almost impossible to properly accredit anything and anyone. Just because someone says something in a vlog it doesn't mean they originated the idea. Hardly anyone acknowledges their intellectual debts on the internet. It's one big theft of thought.

Interviewer: This is all fascinating Jack, and I thank you for bringing it to our attention, but I think we may have strayed a little from the point of our discussion!

Jack: Yes. What was the point?

Interviewer: Why now? Why leave the London Panopticon now for an untested venture?

Jack: Right! Look, I was too young and callow at the time to really understand my role in things. There was no misconduct. No deceit. No intention to mislead. But there was unquestionably a version of this story we could have told that didn't look much like this one. And that got me to thinking of the other times in my career where I went for what was a sellable story as opposed to that which was an authentic reflection of the situation. There were more than I felt comfortable with.

Interviewer: Again, you're not confessing to misconduct?

Jack: No. It's just – there are multiple perspectives from which you can tell a story. Some of those perspectives are more 'true' than others. They're all 'correct' in terms of the facts, but the order and emphasis with which those facts are portrayed can create very different kinds of coverage. Where you start telling a story, and where you end it, is also a powerful framing device. 'She hit me' is a story, but if you go back it might be because 'he hit me first', and so on.

Interviewer: And so, the Centre for Responsible Journalism?

Jack: Yes. This is my attempt to try and put all of this insight into some kind of useful form. There are a lot of media outlets that have offered financial support in exchange for our endorsement of their outlets, and we're working with the journalism department at the University of Alba to build teaching material and professional guidance that others can rely on.

Interviewer: What form is that likely to take?

Jack: Guest lectures, for one thing. We have a lot of journalists signed up to talk about their irresponsible practices. And in some cases their outright misconduct. They get to pass on their reflections to the next generation of journalists. We build those up into professional case studies, and those in turn become available through workshops that we deliver at media organizations.

Interviewer: That sounds great. And what of the students that contacted you about the scandal? Presumably they are delighted to have sparked all this off. Do they get to give a lecture as part of your new program?

Jack: Well… no. When I started digging through the material I realised that I probably had enough for a book, and I pitched that to a publisher who leapt on the idea. So I couldn't actually show those students any of the material in the end because it would compromise the contract I signed. I did have a discussion with them, but I think they were disappointed that it didn't include any new information that they could put in their dissertation.

Interviewer: Thank you so much for taking the time to talk to us today, Jack. Could you tell us a little bit about how we might find out more?

Jack: Yes! If you are a corporation looking to instil a sense of responsible reporting into your journalism, or your research, please get in touch with our finance department who will be able to quote you a price for our bespoke workshops. We have prices beginning at £25,000 for a day's training. My book is much cheaper! It's called 'Jack McKracken and the Ethical News Vendors'. It's being published by Wee Haggis Press – available in September for £25.99 at everywhere good books are sold!'

Interviewer: Jack McKracken, thank you very much and good luck in your new ventures!

OBITUARY: JOHN BLACKBRIAR

John Blackbriar, Esteemed Scottish Academic and Disgraced Researcher, Dies at 63.

Obituary printed in the Dunglen Chronicle

Dr. John Blackbriar OBE, a Scottish academic known for his significant contributions to the field of oil exploration and artificial intelligence, passed away on the 27th of June at the age of 58. For the past two years, he had been battling an especially aggressive form of pancreatic cancer. He died at home, surrounded by his family.

Born and raised in Perth, Blackbriar devoted his life to the study of petrochemistry. After being awarded his doctorate, his work began to explore the economic viability of extracting under-valued resources from the North Sea. His ground-breaking research methods drew accolades and admiration from his peers and the industry alike and generated significant revenue for the Scottish oil industry. It was Blackbriar's pioneering work in the use of artificial intelligence within complex environments that created a mini-Renaissance in the field. His work helped identify dozens of significant untapped and economically viable oil reserves in previously discounted regions of the sea bed. Several oil companies identify Blackbriar's involvement as the thing that kept them from bankruptcy during several of the oil-price crashes in the 90s. One ScotOil colleague one said, 'John's genius was in finding ways to improve return on investment. Any idiot can sink an oil-rig into a huge oil field, but John's work was a multiplier on the profit margin of the whole industry. His work resurrected dying rigs and kept thousands of people employed at times when other parts of the world were laying them off'. This work led to him being awarded an OBE in 2008.

However, in the years leading up to his death Blackbriar's career took an unfortunate turn. Allegations of scientific misconduct emerged, tarnishing his once-unblemished reputation. Despite his previous successes, he faced controversy and eventually chose to resign from his prestigious position at the University of Dunglen. He left behind him a legacy overshadowed by scandal. Blackbriar himself, described by friends as 'a reserved and taciturn man', never commented publicly on the incident which was first reported in this very newspaper.

Blackbriar's former colleague, Sir David Tumblewood, described him as 'An immense talent. An overpowering intellect. And a good friend'. Tumblewood was the principal at the University of Dunglen during the scandal. 'We had our differences, definitely, but today I have lost a good friend and the world has lost a great scholar'.

During his tenure at the university, Blackbriar's work showcased his intellect, passion, and dedication to the pursuit of knowledge. His numerous publications and research findings laid the groundwork for future advancements in the field. His extensive research funding was pivotal in transforming the University from a low-ranked teaching institution to a widely respected research hub. That is, until the reverberations of the BrokenBriar scandal resulted in a brain drain that sent Dunglen back to the tail end of the league tables.

Outside of academia, Blackbriar was known for his love of Scottish culture and tradition. He was an avid bagpiper and enjoyed sharing the melodic sound of his instrument with others. After leaving his professorship, he described himself as enjoying an 'extremely early and unilateral retirement'. He spent most of his time working in his garden and serving as chairman of the local chapter of the Scottish National Party. Colleagues in the party have said that academia's loss was certainly politics' gain.

John Blackbriar leaves behind him a complicated legacy marred by controversy. His contributions to the scientific community and his wider accomplishments will be remembered. They have though become overshadowed by the public disgrace that marked his later years. A complex man, Blackbriar leaves the world considerably less colourful for his passing.

He is survived by his wife, Kate (55), their two daughters – Emily (32) and Lisa (28), and five grandchildren. Funeral arrangements will be private, with a memorial service to be held at a later date. Mourners are asked not to send flowers, but to instead make donations to Macmillan Cancer Support.

CHAPTER 18

Acknowledgements

No one ever writes a book alone, and a book like this – spawned from a series of papers published over many years – has a long list of people to whom we are forever indebted.

We would first of all like to offer our sincerest thanks to Richard Epstein, he of the Case of the Killer Robot that served as the model for this work. He is responsible for sparking a lifelong fascination with ethics and professional conduct. Adding to this, Michael also offers his thanks to Dr Ishbel Duncan. She undoubtedly doesn't remember him as an unbearably cocky student at Abertay University in the late 1990s, but he remembers her and the class in which she introduced this case study to him. This work literally would not exist without either of them.

Michael offers his specific thanks to Mafalda Samuelsson-Gamboa, a wonderful colleague and friend. Her enthusiasm is infectious. Her encouragement unceasing. She is an inspiration and someone who I think every academic should want to be when they grow up. Carrie Morris, who helped me teach this case study at Robert Gordon University, was also instrumental in encouraging the writing of this book – her enthusiasm for the material cut through the world-weary cynicism that produced it. Thanks to Mike Crabb, my partner-in-crime. No specific reason associated with this work, but it's sometimes hard to know where his ideas end and mine begin. It seems weird to acknowledge Pauline Belford, my partner and closest companion, given how she is a co-author of the book. Nonetheless, she is the reason I do pretty much everything I do. I'd probably be a misanthropic hermit throwing beer bottles at pigeons were I not perpetually trying to make her proud.

DOI: 10.1201/9781003426172-18

Pauline would like to thank Michael Heron. He is my co-author, but he is so much more than that. He is my partner, my best friend, and my most supportive work colleague. He is my biggest cheerleader, and I'm not sure where I would be in life without him by my side. Actually, I am pretty sure I would still be living in Dundee, not having been brave enough to move and take advantage of the opportunities academia gives you to explore the wider world. Michael has challenged me in good ways, and my life has been much richer and more fulfilling as a consequence.

The original form of this book was a series of papers published in ACM's Computers and Society journal. In the ten years between the first and last paper, this work was supported by Dee Weikle and Richard Blumenthal in their role as editors. It is a rare and precious indulgence for them to permit room for a couple of authors to spend years exploring the implications of a single case study. We are forever indebted to them. We are immensely thankful to everyone at Taylor & Francis – especially Elliott Morsia – for taking us through the journey of seeing our work in a bookcase as opposed to a digital library.

We are thankful too to the hundreds of students we have taken through the Scandal in Academia, and for their thoughts, opinions, and discussions. The work is richer for the many insights you all sent our way. Michael is appreciative too of Dr Roger McDermott, head of Undergraduate Computing at Robert Gordon University. Upon being informed by a worried colleague regarding the often sharply critical tone associated with this case study's deconstruction of academia, Roger merely shrugged and said, 'I'm sure Michael knows what he's doing'. Thanks for the vote of confidence when I was **so obviously** a fifth columnist attempting to bring the system down from the inside.

Thanks to our parents for being forever baffled by the way academia works, and yet supporting us as we work our way through it.

Finally, thanks to those of you who have bought and read this book. Of all the things you could have been doing with your time, you decided to do that. You have our deepest gratitude.

Index

D

E

T

X

Y

Z

Printed in the United States
by Baker & Taylor Publisher Services

Printed in the United States
by Baker & Taylor Publisher Services